FOREWORD

The several hundred anole species of the genus *Anolis* are found in the New World from southern North America to Paraguay and Bolivia. They are particularly well represented in the Caribbean, where virtually every island harbors different species.

For decades anoles have been among the most popular charges in our terraria. Unfortunately, published accounts of these lizards as a rule are limited to the most familiar species, such as the Green Anole (*Anolis carolinensis*) and the Knight Anole (*A. equestris*). Other anoles worthy of mention are discussed either incompletely or not at all.

This volume is devoted strictly to the "C

keepi

in our terraria and have been able to study various species in their natural homes in the Caribbean. Here we treat about 20 species, some with numerous subspecies, that are the most likely to be noticed or collected by interested vacationers and terrarium keepers. Many are colorful, others are drab, but all can give pleasure to their observers or keepers. Anyone with an interest in caring for and breeding anoles will discover relevant information in this volume on managing anoles in the terrarium.

Ralf Heselhaus and
Matthias Schmidt, Muenster
Translated by William Charlton

PHOTO BY JIM MERLI.

There are relatively few Caribbean anoles available in the herpetocultural trade. One that does turn up is the Knight Anole, *Anolis equestris*, which grows to well over 40 cm/16 in in length and does quite well in captivity.

SYSTEMATIC CONSIDERATIONS

The most beautiful and interesting lizards of the New World are members of the iguanid family or superfamily Iguanidae or Iguanoidea, depending on your taxonomy. (We use the "old" classification of a single family here, but the iguanoids were broken into several families by Frost and Etheridge in 1988; for a brief recap, see the article by Walls in *Tropical Fish Hobbyist*, Dec., 1991.) With about 700 species in more than 50 genera, it is one of the most species-rich lizard families. The large species, like the familiar Green Iguana (*Iguana iguana*), with a total length of up to 2 meters, and the almost equally large Rhinoceros Iguana (*Cyclura cornuta*), resemble superficially the dinosaurs of bygone eras in Earth's history.

The iguanids, which are found mainly in the New World, are scarcely distinguishable externally from the agamid lizards of the Old World. The safest distinguishing characteristic is the arrangement of the teeth. With the iguanids the teeth are located on the inner side of the jawbone; with the agamids, on top of it.

Within the Iguanidae, usually five subfamilies are distinguished. The subfamily Anolinae includes, besides the anoles, the following genera: *Anisolepis, Aptycholaemus, Chamaeleolis, Chamaelinorops, Cupriguanus, Diplolaemus, Enyalius, Leiosaurus, Phenacosaurus, Phrynosaura, Polychroides, Polychrus, Tropidodactylus, Urostrophus*, and

ARTWORK BY JOHN R. QUINN.

Chamaelinorops barbouri.

PHOTO BY RALF HESELHAUS.

A courting male Green Anole, *Anolis carolinensis*.

ARTWORK BY JOHN R. QUINN.

Chamaeleolis chamaeleonides.

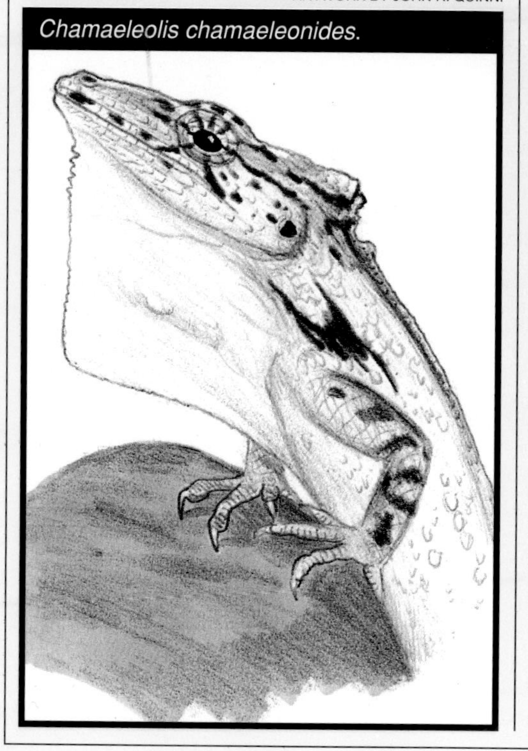

Xiphocercus. (Taxonomically inclined hobbyists should note that under the Frost and Etheridge system the anoles fall into the family Polychridae, an older name than Anolinae. Additionally, many specialists now split *Anolis* into a variety of genera that are not recognized here, especially the full genus *Norops*.)

The anoles, with about 300 species and almost as many subspecies, are the largest genus of iguanids. Aside from a few large species such as the Knight Anole, *Anolis equestris*, and the False Chameleon, *Chamaeleolis chamaeleonides*, which can reach a length of 50 and 35 cm (20 and 14 inches), respectively, as a rule anoles are small to medium-sized lizards. As tree-dwellers, most species are characterized by a streamlined body, a long tail, and robust legs with adhesive pads on the toes.

PHOTOS BY RALF HESELHAUS (TOP LEFT), MATTHIAS SCHMIDT (TOP RIGHT), MATTHIAS SCHMIDT (BOTTOM).

Above: Brown Anole, *Anolis sagrei*. **Right**: *Anolis trinitatis*. **Below**: *Anolis oculatus winstoni*.

WHERE DO ANOLES LIVE?

The range of the anoles extends from the southern United States across Central America and the Caribbean islands to Paraguay and Bolivia in tropical South America.

About 100 species, many with several subspecies, live on the Caribbean islands. This abundance of species resulted from the changes to the Caribbean island world following the Ice Ages. The rising water levels flooded many islands, so that finally only the mountains of the once-large islands were retained as land masses. Through the fragmentation of the formerly continuous living space, the original lizard fauna also was fragmented. Separated from one another as island forms, the lizards developed in the course of evolution into new species with special requirements and capabilities.

On the large islands, the varied landscape and plant communities made possible the formation of numerous anole species specialized for certain microhabitats or diets; these could live side by side without interference. On Cuba we find about 40 species that are distributed through various ecological niches. The giant anoles, such as the Knight Anole, are found in the tops of the trees in the rainforest. Because of their size, they manage food animals that are too large for smaller anoles to handle.

PHOTO BY MATTHIAS SCHMIDT.

Biotope of *Anolis ferreus* on Marie-Galante.

Total geographic range of the anoles.

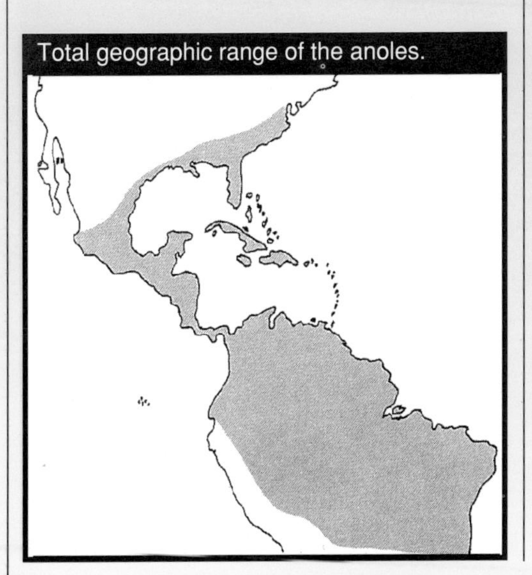

The arboreal anoles, such as *Anolis porcatus*, are medium sized and live in the trunk region of the trees, in taller shrubs, or on houses. Finally, the grass anoles such as *Anolis alutaceus* live in low shrubs in fields and along roads. As a rule, they are scarcely longer than

10 cm (4 in) and feed on small insects. Of the Cuban anoles, Peters, an expert in the group, in 1970 distinguished further the group of the bark anoles, *Anolis loysiana*, and the gecko anoles, *Anolis bartschi*. The gecko anoles, in contrast to most of the other species, live in wet and shaded places, such as in caves and crevices.

PHOTO BY RALF HESELHAUS.

Right: The Brown Anole, *Anolis sagrei*, is lord of its domain on Isla de la Juventud, Cuba. **Below:** Occupation of various ecological niches used by certain Cuban anoles.

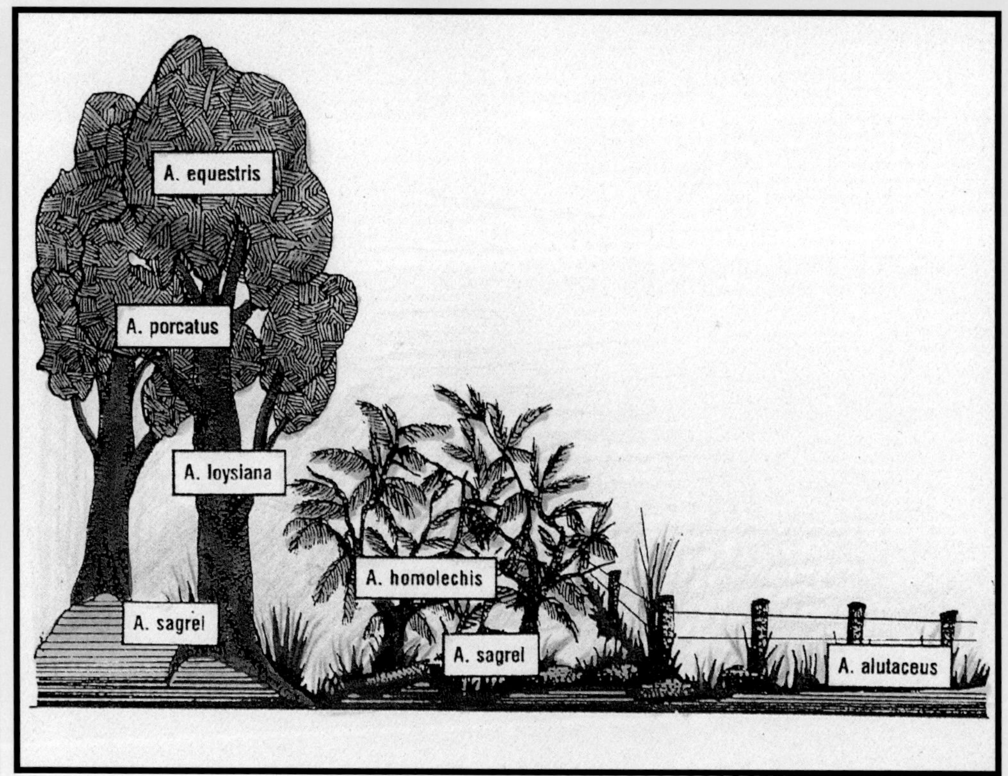

SPECIAL FEATURES

Anoles are provided with several special features that are not present in most other iguanids. They seem to have been equipped preferentially by nature; not only do they have a streamlined build and bright colors, they also exhibit characters that otherwise are confined to other lizard families. For example, anoles, like the geckos, have adhesive pads on their feet and, like the chameleons, can change their color.

The acrobatic life in the trees demands yet another specialization of the foot anatomy. To largely eliminate the possibility of falling from airy heights, anoles work with double "safety nets." The middle toes are enlarged and carry on the underside transverse lamellae that provide a secure hold on a smooth substrate. On the ends of the toes there are claws to provide a hold on uneven surfaces. With their unique climbing feet, anoles can move effortlessly and safely on both smooth leaves and knotty tree bark.

PHOTO BY GUIDO DINGERKUS.

The colorful dewlap often is present on both male and female *Anolis* specimens. This male has a particularly pretty throat fan.

PHOTO BY RALF HESELHAUS.

One of the more attractive Caribbean anoles is *Anolis trinitatis*. Here a male scales a pane of glass by making use of the enlarged middle toes, which have transverse lamellae on the underside.

The little anoles have also stolen something from the chameleons. Some species, like *Anolis carolinensis* and *Anolis porcatus*, are capable of an amazing change in color. The lizards can change their color in a matter of minutes from light green to dark brown. This ability is the reason why the Green Anole also is called the "American Chameleon."

Finally, one of the most conspicuous characteristics is the throat fan or dewlap that often is present in both sexes. In males it sometimes attains an astounding size and can extend well behind the front legs. Its signaling effect is increased by bright colors. When excited, anoles unfurl the dewlap with the aid of the elongated hyoid cartilage and erect it like a fan, accompanied by vigorous head bobbing. The colorful fan is used during mating and territorial defense.

MASTERS OF THEIR DOMAIN

With rhythmic head bobbing and erect throat fan, a male *Anolis sagrei* threatens its neighbor: Don't come closer or there will be trouble. The territorial fights of the males as a rule are carried out from a distance. It is a question here of ritual fights in which the males, through ceremonial behaviors, stake their territorial claims.

When two males battle at a distance of several meters with head bobbing and swaying threat behavior as well as hoisting the brightly colored throat fan, they actually are fighting over imaginary territorial boundaries. After a while one of the males withdraws slowly as the threatening behavior fades away. The test of strength from a distance ends without the defeated rival coming to harm.

In the terrarium the territorial disputes do not always end so benignly. The defeated male often does not have the opportunity to get away from the victorious territory owner. He is subject to stress, is constantly ready to take flight, and finally even stops feeding. But we must never let things get this far!

When keeping anoles together, it must be kept in mind that normally we can only keep one pair in the terrarium. In larger terraria it also is possible to keep one male and several females. Close observation of all individuals soon shows whether any are being oppressed. If this is the case, we must remove from the terrarium any anole that is being picked on.

If we want to keep together several species in a large terrarium, we should choose species of equal size but different coloration. We have had good experiences with the communal care of *Anolis sagrei* and *Anolis marmoratus speciosus*. It has been reported that a pair of the "flat-headed," generally green species such as *Anolis carolinensis* and *Anolis porcatus* can be kept successfully with a pair of "high-headed," usually brown species such as *Anolis sagrei, Anolis sabanus*, or *Anolis lineatopus*. This could be due to the different territorial requirements and the different activity periods of the species. Whereas *Anolis porcatus* and *Anolis carolinensis* are purely diurnal and inhabit the upper levels of trees and shrubs, the more crepuscular *Anolis lineatopus* stays predominantly in the lower trunk region.

Now there may well be terrarium keepers who claim to keep six anoles of three species in their 60-cm (2-foot) cage and have no problems at all with aggression. This may even be true, since anoles kept under these kinds of cramped conditions cannot form territories at all and therefore do not exhibit any territorial behavior. This overpopulation, however, is completely out of place with appropriate keeping. Here less would be more, particularly since anoles exhibit the full range of their interesting behavior only under suitable environmental conditions.

Finally, other terrarium animals also can be kept in the anole terrarium. An association with frogs and toads or with day geckos is possible. Naturally, the demands of the other terrarium animals must also be considered. This sort of association for ornamental purposes alone, however, should be rejected by the conscientious terrarium keeper for reasons of species protection.

Although many male Caribbean anoles have strong territorial instincts, some, like this Brown Anole, *Anolis sagrei*, seem more amenable to sharing their captive quarters. As a rule, however, you should get into the habit of placing only one male into each enclosure (unless, of course, the enclosure has a great deal of space). Photo by R. D. Bartlett.

COURTSHIP, BREEDING, AND REARING

For anoles, as all other animals, reproduction is one of the most important tasks in life. During the breeding season, which lasts several months in many species, the male anole uses all of his physical attributes to impress the ladies. When he spots a potential mate, the male shows his best side. With body raised high, he presents his colorful throat fan, and vigorous head bobbing betrays his inner excitement.

When the female tries to evade the male's advances, the male frequently chases her. The male anole eventually catches his mate and tries to grab the female's nape region with his mouth. If he is energetically shaken off by the female, he backs off. If the female indicates her readiness to mate by tolerating the male's advances, the nape biting soon is followed by copulation. In the following weeks the female becomes increasingly plump, and a few weeks after copulation she seeks out a suitable place on the ground to lay her clutch of one or two eggs. The eggs are buried, for example, in moist soil in a small hollow the female excavates with her front legs. Because the female smooths the surface over, the site of egg-laying sometimes remains hidden from the keeper.

PHOTO BY R. ALLAN WINSTEL.

A mating pair of Green Anoles, *Anolis carolinensis*, in Silver Springs, Florida.

PHOTO BY GUIDO DINGERKUS.

Anole eggs are remarkably tiny, and clutch sizes usually number only one or two.

We can find many clutches, however, if we carefully examine suitable sites in the terrarium. The first indication that we should start our search is the presence of a slender female again. When the eggs are found, they should be taken from the terrarium and transferred to a prepared incubator. In so doing we must make sure not to change the position of the eggs, since in some cases this can lead to the death of the

BOTH PHOTOS BY MICHAEL GILROY.

embryos. Small transparent plastic boxes with ventilation slits in the cover, such as are available in some pet shops, make suitable incubation containers. Fill a box or small aquarium a fourth full of water and place a brick on the bottom, the surface sticking out of the water. On top of the brick place the incubator with the eggs. Cover the incubator with a pane of glass placed at a slant so that water cannot drip on the eggs.

The incubator itself consists of a small hard plastic box filled with about 5 cm (2 in) of substrate. A suitable substrate is a mixture of peat and sand or, better yet, vermiculite, which can retain moisture for a very long time and release it slowly. The substrate should be moistened to saturation without creating standing water. The eggs are put in small hollows such that the top side is above the substrate. During the period of maturation, a temperature of 25 to 28°C (77 to 82°F) should be maintained in the incubator. Be sure the substrate remains moist throughout incubation.

Once a young anole has hatched, an exciting period of rearing begins for the terrarium keeper. The initially tiny anole must be well fed. Freshly

Above: Getting anoles to reproduce in captivity really isn't all that difficult, but you must pay close attention to their climatic requirements. Changes in photoperiod, temperature, and humidity all play a role in triggering reproductive behavior. **Below:** Note the sexual characteristics of this male Green Anole, *Anolis carolinensis*—the postanal pores and the paired hemipenal bulges.

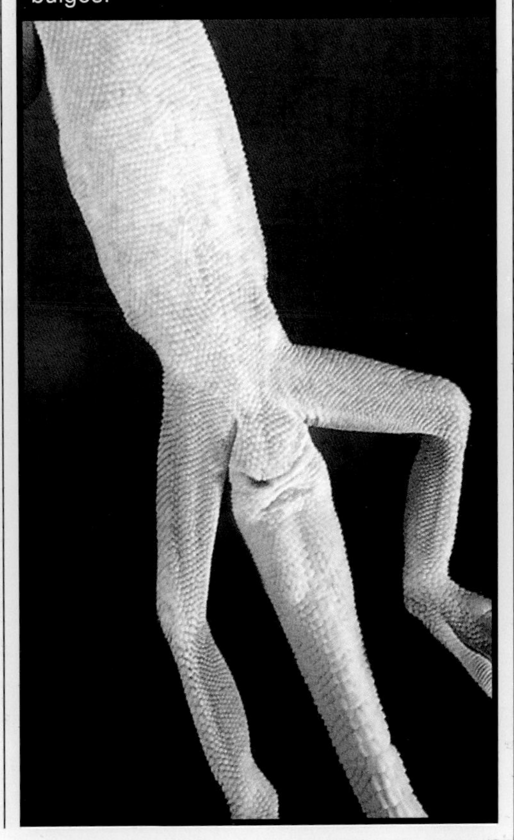

hatched Green Anoles are only 5 to 6 cm (2 to 2.5 in) long. The youngsters are placed individually in prepared small terraria, such as plastic containers with ventilation surfaces in the cover and on one side. The "solitary confinement" is necessary because of the aggressiveness

PHOTOS BY RALF HESELHAUS (TOP) AND PAUL FREED (BOTTOM).

Left: *Anolis marmoratus speciosus* is a remarkably hardy terrarium subject that will breed willingly. Most specimens are fairly attractive, but unfortunately they do not show up in the herpetocultural hobby too often. **Below:** An unusual color pattern in a Green Anole, *Anolis carolinensis*, which normally is uniformly green or brownish on the back.

present even in young anoles. The rearing cage can have spartan furnishings, but small twigs and a few shoots of *Scindapsus* should be present to serve as climbing facilities and shelter for the baby anoles. At first a piece of foam rubber, cut to fit the bottom and kept constantly moist, makes a suitable substrate.

Youngsters of the small species are fed at first with *Drosophila* (fruitflies) and freshly hatched crickets. With the larger species, like the Knight Anole, the youngsters are already 12 cm (5 in) long on hatching and manage medium-sized insects from the start. Some sweets (honey) are taken readily by young anoles. Drinking water is sprayed daily on a leaf or on a side of the rearing cage, without spraying the anoles in the process.

The food animals must be dusted regularly with a vitamin-calcium preparation. Many brands are available at your pet shop. These preparations can safely be administered daily. A daily illumination of the young anoles with full-spectrum light of the proper type at the correct distance (see your pet shop) is recommended but is not absolutely necessary if adequate amounts of vitamins are provided.

The young anoles should not be kept too cold at night. Not letting the temperature at night fall below 25°C (77°F) prevents deadly colics or cramps in the youngsters. The cramps may result from poorly digested food at too-low temperatures, as we have been able to confirm in our own terraria.

With appropriate attentiveness, it will be easy to rear healthy and robust anoles that can in turn produce offspring when they reach sexual maturity.

PHOTO BY ISABELLE FRANCAIS, COURTESY OF MARK B. ROBERTSON.

Males generally should be kept separate from other males, but one male can be kept with three or four females. Shown is an adult male/female pair.

ANOLE SPECIES

PHOTO BY RALF HESELHAUS.

The Cuban Grass Anole, *Anolis alutaceus*, is widely distributed in Cuba. It needs a lot of terrarium space and good ventilation.

Anolis alutaceus, Cuban Grass Anole

The "Sugar Island" of Cuba is home to at least 41 species of anoles and is a center of distribution for the Caribbean anoles. This abundance of species probably is the result of the size of the island and its diversity of landforms. With a length of 1200 km (720 mi), Cuba is the largest Caribbean island. In the high mountains (up to 2000 m, 6600 ft) we find a variety of zones of vegetation. In the western part of the island a hot and humid tropical climate prevails, with abundant rainfall in the summer months, whereas the east during the dry season exhibits more of a savannah climate.

Besides the main island, Cuba consists of a large number of smaller islands, the majority of which are uninhabited by man. Isla de la Juventud, the "island of youth" (formerly Isle of Pines), is the second largest island but is only 40 km (24 mi) long.

Description: *Anolis alutaceus*, with a total length of 10 to 11 cm (4 in), is one of the smallest anoles. The very slender anoles, brown with a sprinkling of light spots, display a light stripe on the side. Another faint stripe extends from the lips to the external ear opening. The ventral side is beige-white. In contrast to this plain coloration is the rather large throat fan with its bright yellow color.

Range and habitat: *Anolis alutaceus* is widely distributed on Cuba. The little iguanids are abundant on the Isla de la Juventud, where they perch on thin twigs in low bushes along the sides of the road. Because of their brown ground color, they are hard to detect at first, particularly since when they are approached they slide around the twig and become "invisible."

Care in the terrarium: Despite the small size of *Anolis alutaceus*, we must not choose too small a terrarium and should offer a height and width of at least 60 cm (24 in). Because these anoles are quite sensitive to stuffy air, the terrarium must be equipped with sufficiently large ventilation openings.

The recommended furnishings include, besides thin branches, a planting with small palms, long-leafed bromeliads, and *Scindapsus*. The temperatures should reach 30°C (86°F) locally (spotlight); at night the temperature can drop to 22°C (72°F). Through daily spraying the humidity is raised to 90 percent temporarily, but during the day the terrarium must be able to dry out completely again.

Anolis alutaceus should be fed a varied assortment of vitamin-enhanced small insects. Regular exposure to ultraviolet radiation is beneficial to the anoles' health.

Anolis bartschi, Cuban Cave Anole

Description: *Anolis bartschi*, with a total length of 27 cm (10.8 in), is one of the medium-sized anoles; females, at a length of about 20 cm (8 in), are distinctly smaller than males. There are brown, gray, creamy yellow, and pale blue varieties; the anoles are not exactly colorful, but they are very attractively colored. In addition to the brown ground color, *Anolis bartschi* displays gray saddle-shaped blotches on the back and scattered cream-yellow, irregular vertical stripes on the sides. *Anolis bartschi* is at its most beautiful when excited and in subdued light. The lower eyelids and adjoining head region up to the ear opening then are a splendid bluish violet, and the remaining markings on the body also take on a greater contrast.

The females are altogether plainer; on the gray-brown dorsal coloration they display three dirty white, repeatedly interrupted longitudinal stripes, one in the middle of the back and two others on the sides. Juveniles are the same color as females until they reach sexual maturity.

The skin appears translucent and velvety. The throat fan of *Anolis bartschi* is conspicuously small. When erected it is only half the size of the head.

Range and habitat: *Anolis bartschi* occurs exclusively in western Cuba in the province of Pinar del Rio. It is found there in caves and crevices in the limestone prevalent in the biotope, and into which the anoles withdraw at higher temperatures. The cave anoles primarily hunt at night. In the terrarium, on the other hand, they prove to be diurnal.

PHOTO BY RALF HESELHAUS.

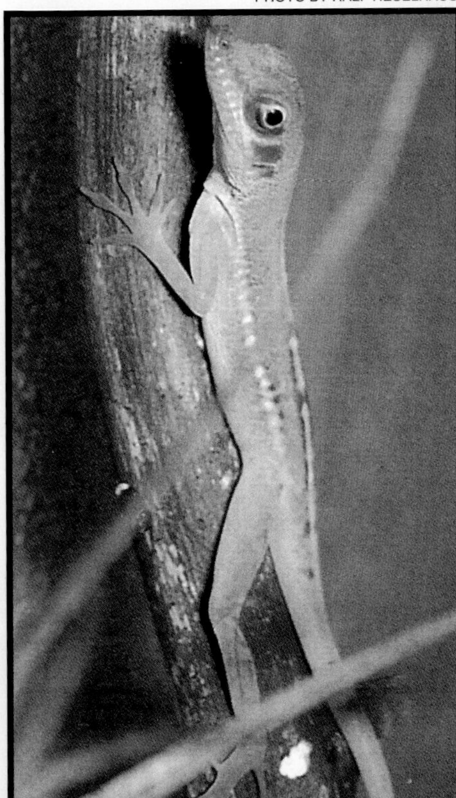

The Cuban Cave Anole, *Anolis bartschi*, can be found only in the western Cuban province of Pinar del Rio. It needs a relatively high humidity and a rock wall.

Care in the terrarium: The specific requirements of *Anolis bartschi* are met in the terrarium by trying to mimic a rocky cave wall. Suitable rocks like limestone or dolomite are layered against the back wall, resulting in a fissured rock wall. Branches and plants in the foreground provide additional places to climb and act as visual barriers.

The substrate, a mixture of peat and sand, should be kept constantly moist to achieve a sufficiently high relative humidity of 75 to 85 percent. The anoles can obtain their drinking water through daily spraying. The diet consists of the usual insect food, which should be fortified regularly with minerals and vitamins.

The successful breeding of *Anolis bartschi* has been reported. The anoles were kept in pairs in a 65 X 50 X 30 cm (26 X 20 X 12 in) glass terrarium that was illuminated by two fluorescent bulbs for 11 to 16 hours as well as with a 20-watt UV-A fluorescent tube for about three hours a day. The temperature varied between 20 and 28°C (68 and 82°F) and was up to 35°C (95°F) in the vicinity of an incandescent bulb hanging in the terrarium.

PHOTO COURTESY OF FOUR PAWS.

Organic substrates like crushed bark are ideal for anoles. Most are easy to work with, pleasing to the eye, and can be bought in bulk quantities.

The cue for the start of reproductive activity in May and June may be exposure to sunshine in the spring. The courtship behavior is not very pronounced, and copulation follows immediately after prior intervals of head bobbing. Because the males chase the females actively during the mating season, it is advisable to separate pregnant females from the males. Keeping this species with other anoles, such as *Anolis sagrei, Anolis luciae, Anolis distichus*, or *Anolis porcatus*, seems beneficial because the ready-to-mate males can be diverted in this manner, which is to the benefit of the females.

The clutch, which usually consists of a single egg, in the terrarium was covered only slightly with peat under volcanic limestone. At temperatures between 22 and 28°C (72 and 82°F), the young hatched in the incubator after 60 to 70 days. The young were reared individually in small, planted terraria. The young anoles took young crickets and waxworms fortified with vitamins and minerals. The captive-bred anoles were sexually mature at 12 to 15 months.

Anolis bimaculatus bimaculatus

Description: *Anolis bimaculatus bimaculatus* is a very attractively colored anole. The dorsal side is grassy green to moss green with a pattern of black dots or specks. From the snout to the ear opening extends a yellow stripe that continues slightly displaced beyond the shoulder and ends abruptly at the height of the elbow. The male exhibits an erectile crest in the occipital region that continues as a low, non-serrated dorsal crest. The male displays the occipital and dorsal crests, as well as the green or yellow, white-spotted throat fan, only when excited. Adult *Anolis bimaculatus bimaculatus* grow to a size of 30 cm (12 in) or more in the males; females reach a length of only 20 cm (8 in).

Range and habitat: *Anolis bimaculatus bimaculatus* occurs in the Lesser Antilles (St. Christopher, St. Eustatius, Nevis). As an arboreal species, the anole primarily inhabits

PHOTO BY DAVID GREEN.

Anolis bimaculatus is native only to the Lesser Antilles, where it lives in tall palm trees. To keep the species you need a tall terrarium and plenty of plants, such as yuccas.

PHOTO BY RALF HESELHAUS.

tall palms in secondary forest.

Anolis gingivinus and *Anolis sabanus*, often considered subspecies of *bimaculatus* in the older literature, were raised to species status by specialists in Caribbean anole taxonomy. We agree with this interpretation, so the only remaining subspecies recognized is the brown- to gray-marked *Anolis bimaculatus leachi*, which is native to Antigua and Barbuda.

Care in the terrarium: We should take into account the need of *Anolis bimaculatus bimaculatus* for a life in airy heights. In the terrarium, a cage height of at least one meter and an appropriate furnishing with tall-growing plants, such as yuccas, is adequate. The anoles prefer a well-lighted and well-ventilated terrarium with temperatures of 30°C (86°F) during the day (about 23°C, 73°F at night). Through daily spraying we achieve the required high humidity and meet the water requirements of the anoles. *Anolis bimaculatus bimaculatus* readily accepts green insects as well as smaller lizards!

Anolis carolinensis, Green Anole

Description: The most familiar anole is doubtless the Green Anole. Whole generations of terrarium keepers have enjoyed this pretty, graceful arboreal lizard and have also bred it regularly. The Green Anole displays a bright green coloration that is interrupted only by a light, jagged dorsal stripe or subdued small spots in some individuals. The ability to change in seconds from green to brown with a change in mood has given the Green Anole the name "American Chameleon." Males possess a pink throat fan with white scales. The fan is extended when the anole is excited, such as during courtship or in territorial defense.

The Green Anole, *Anolis carolinensis*, is one of the most popular lizard pets in the world. Photo by Michael Gilroy.

Males grow up to 22 cm (8.8 in) long; females are slightly smaller.

Range and habitat: *Anolis carolinensis* occurs primarily in the southeastern United States, and only at the southern limit of the range does the species nearly reach the Caribbean realm, occurring in the Florida Keys. Related species (once considered subspecies or introduced populations of the Green Anole) occur in the Bahamas and Cuba. Green Anoles inhabit bushes and deciduous trees and, as followers of civilization, are also found in houses, stables, and gardens.

Care in the terrarium: *Anolis carolinensis* should be kept in pairs—or one male with two females—in medium-sized terraria of about 80 X 50 X 50 cm (height X width X depth), 32 X 20 X 20 in. Specimens from the southeastern United States need temperatures between 20 and 28°C (68 and 82°F) and moderate humidity. Green Anoles from the Florida Keys can be kept at higher humidity and higher temperatures.

Particularly for breeding, a two-month hibernation at temperatures between 15 and 20°C (59 and 68°F) is necessary. The anoles eat little during this time, but must be provided regularly with water.

The breeding season begins as a rule a few weeks after the end of hibernation. The male approaches the female with fully extended throat fan and announces his intentions with vigorous head bobbing. After the male has bitten the female's nape, copulation takes place.

The pregnancy of the female lasts two to three weeks. Then she buries her clutch of two eggs in damp soil. If you are able to find the laying site, transfer the clutch, taking care to maintain the eggs' original

PHOTO BY R. T. ZAPPALORTI.

Occurring primarily in the southeastern United States, Green Anoles, *Anolis carolinensis*, are best kept either in pairs or in sets of three (one male, two females).

orientation, to an incubator. The young, which are 6 cm (2.4 in) long at first, hatch after about 40 days. We put them individually in small rearing cages. The baby anoles are fed *Drosophila*, small waxworms, and other small insects that we dust regularly with a vitamin preparation.

PHOTO COURTESY OF FOUR PAWS.

Sheets of artificial turf (also known as indoor/outdoor carpeting) make good substrates for Caribbean anoles. Such sheets come in a variety of sizes to fit a variety of enclosures and can be used over and over.

Anolis cuvieri, Puerto Rican Giant Anole

Description: *Anolis cuvieri*, along with *Anolis equestris* and *Anolis ricordi*, ranks among the giants of its genus. Males reach a total length of almost 40 cm (16 in); females remain slightly smaller.

This impressive anole is a handsome light green; only when it is out of sorts does the color change to an olive green with pale spots. The massive head with well-developed jaw musculature indicates that the giant anole can deliver a hard bite. The big eyes are covered by attractive eyelids with black and green markings. The throat fan, which is present in both sexes, in the erect state attains twice the size of the head in the male. The throat fan is yellow in the male, yellow-green in the female.

Both sexes exhibit a serrated dorsal crest that is more strongly developed in the male. Males also exhibit a wavy fold of skin on the tail.

Range and habitat: *Anolis cuvieri* comes from Puerto Rico and the neighboring Virgin Islands. There they inhabit the trunk and crown region of large trees at a height of several meters. The preferred biotope is shade trees in coffee plantations, but today they also are found in large trees around hotels.

Care in the terrarium: *Anolis cuvieri* needs roomy, well-ventilated cages starting at a height of 1 meter (39 in). The humidity should be 70-80 percent and the temperature about 28°C (82°F) during the day, with a slight nocturnal cooling. The

Native to Puerto Rico and the neighboring Virgin Islands, the Puerto Rican Giant Anole, *Anolis cuvieri*, needs a fairly high terrarium with plenty of slanting, leafy branches.

PHOTO BY RALF HESELHAUS.

furnishings consist of slanting branches shaded by large-leafed plants and pieces of bark. Besides large insects, we can also feed vitamin-enriched beef heart in order to satisfy the voracious lizards. A 12-cm (4.8-in) juvenile hatched in the terrarium following an incubation period of 95 days at temperatures of 20 to 25°C (68 to 77°F). The initially brown youngster grew to 19 cm (7.6 in) in 15 weeks and slowly took on the green adult coloration.

Anolis equestris, Knight Anole

Description: The Knight Anole, with a total length of more than 40 cm (16 in), is one of the giants among the anoles. Because of its impressive appearance and relative ease of keeping, it has been a permanent guest in our terraria for many years.

The massive, strongly compressed "alligator head" and the robust legs with sturdy claws identify the Knight Anole as a tree-dweller. Currently eight subspecies are recognized, but there are indications that other subspecies may be worth naming. The subspecies differ primarily in the

The familiar Knight Anole, *Anolis equestris,* is, at a length of more than 40 cm/16 in, one of the giants among the anoles.

PHOTO BY RALF HESELHAUS.

coloration of the throat fan, although it has been pointed out that specimens with differently colored throat fans can occur in the same population. Because females also possess throat fans, the sexes can be distinguished only on the basis of the

PHOTO COURTESY OF HAGEN.

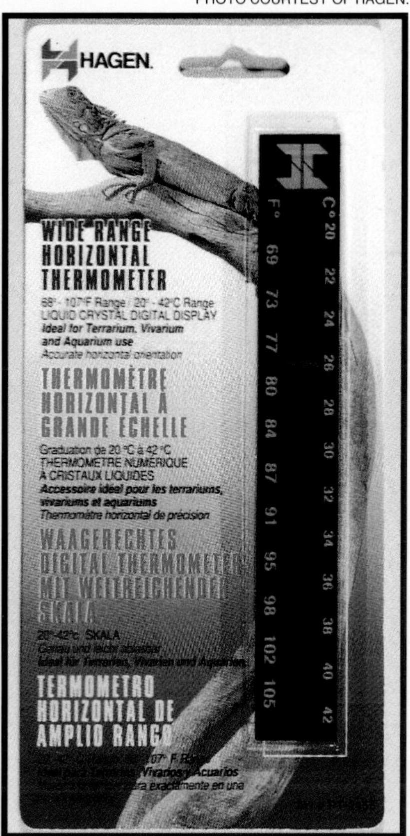

Keeping an eye on the temperature in your anole's terrarium is an important facet of good husbandry. If the animal is allowed to become too warm or too cold, it could become ill. Fortunately, high-range thermometers designed specifically for herp-keeping are now available.

thickened base of the tail and the broader head of the male.

As a rule, Knight Anoles exhibit a bright green to dark green dorsal coloration. In many specimens the skin between the scales shows

The Knight Anole, *Anolis equestris*, is native to Cuba, but there now are thriving introduced populations in Florida as well. Photo by Mella Panzella.

through as yellow or light green, so that the coloration can appear mottled. The eyes have a dark mask. From the base of the front leg runs a light yellow lateral stripe that ends before the middle of the body. Males in particular exhibit a low, serrated dorsal crest.

kept sufficiently moist and the cage is sprayed with water once a day. Because *Anolis equestris* drinks a lot, we must take this requirement into account and offer the Knight Anoles lots of water, if necessary by means of a pipette. The terrarium should be well ventilated. In terraria with little air movement, temporary ventilation can be achieved with a small fan. In the upper parts of the terrarium, with the aid of fluorescent tubes and a spotlight, the required temperatures of 28 to 30°C (82 to 86°F) are easily achieved. At night the terrarium can cool to about 20°C (68°F).

PHOTO BY R. D. BARTLETT.

A young Knight Anole, *Anolis equestris*, from an introduced population in Dade County, Florida.

Range and habitat: The Knight Anole has been introduced into Florida from its original range in Cuba. The U.S. population scarcely differs from the Cuban populations. As an unexcelled tree-dweller, the Knight Anole lives in the tops of tall trees. Juveniles, however, occasionally are found in the lower parts of trees and in tall shrubs.

Care in the terrarium: For the large Knight Anoles, despite their very sluggish habits, only roomy terraria starting at a height of 100 cm (39 in) will suffice. The furnishings should consist of sturdy branches partially covered by large-leafed plants, such as philodendrons, to provide hiding places for the anoles. A water dish can be dispensed with if the substrate is

Despite the relatively simple care, the Knight Anole seldom has been bred in the terrarium, though some success with specimens kept in greenhouses has been achieved. Successful breeders have fed their anoles with various insects, baby mice, earthworms, and beef heart once or twice a week. The food animals were enriched with minerals and vitamins once a week.

PHOTO COURTESY OF HAGEN.

Under-tank heating pads work well with most Caribbean anoles, though the terrestrial forms will benefit more than the exclusively arboreal species. A keeper can warm one particular section of the enclosure, giving the inmate more than one temperature zone to choose from.

The clutch of one or two eggs is buried in the substrate. One breeder placed the eggs in fibrous peat moss and matured them in an incubator at temperatures of 18 to 24°C (65 to 75°F). The youngsters, initially 12 cm (4.8 in) long, hatched after 72 to 92 days. Rear the young individually and give them large doses of vitamins and calcium. After about 15 to 18 months, the young Knight Anoles reach sexual maturity.

Anolis extremus

Description: The males, approximately 20 cm (8 in) long, are a rich green on the legs and tail. The head area is dark blue to black, and the back is a pale blue-gray on which numerous dark blue to black specks are distributed. The throat fan is dark orange and exhibits small black spots. The females (about 12 cm, 4.8 in) are brownish green and exhibit dark brown spotted or striped markings and a light dorsal stripe.

Range and habitat: *Anolis extremus* is native to the island of Barbados, where it is abundant in many different habitats. It was introduced from Barbados to St. Lucia and is found there primarily in the western part of the island. The anoles are particularly abundant in the vicinity of the capital of Castris and its environs. *Anolis extremus* lives on St. Lucia in slightly moist biotopes and prefers large, thick hedges and trees, where the anoles like to seek out sunny places.

Care in the terrarium: Large, well-ventilated, and well-lighted terraria are suitable for keeping *Anolis extremus*. The anoles need sturdy climbing branches and foliage that is not too thick, so that both basking sites and cooler places in the shade are present. Daytime temperatures of 30°C (86°F) and a humidity of up to 85 percent are preferred by these anoles.

The males of *Anolis extremus* grow to around 20 cm/8 in, which is almost twice as long as the females (12 cm/4.8 in). This species needs both warm and cool spots in the terrarium and a humidity of around 85%.

PHOTO BY R. D. BARTLETT.

Anolis garmani, Jamaican Anole

Description: *Anolis garmani* is one of the larger anole species. Males attain a total length of up to 35 cm (14 in); at 26 cm (10.4 in), the females are clearly smaller. The Jamaican Anole is a pretty leaf-green on the dorsal side; the belly is white. When excited the anoles display brown vertical stripes in the green dorsal coloration. At night the dorsal side darkens to brown.

Males possess a large, yellow throat fan and a serrated crest that begins with tall spikes on the nape, becomes lower on the back, and gets higher again on the tail. In contrast to the females, the males can be distinguished by the enlarged postanal scales.

Range and habitat: *Anolis garmani* lives on Jamaica. In the lowlands the anoles live in isolated, tall trees and bamboo, whereas in uplands they are found in forests. The species has been introduced into southern Florida.

Care in the terrarium: On account of its size and at times pronounced nervousness, *Anolis garmani* needs roomy terraria with a height of at least 100 cm (39 in). As furnishings we can use a large rootstock of resinous bog wood, which with its bizarre branching offers the anoles ample opportunity to climb. A cork cylinder reaching to the top of the terrarium and large-leafed plants like philodendrons round out the furnishings. The air temperature should be about 30°C (86°F) during the day and 22°C (72°F) at night, with a humidity of about 80 percent.

Anolis garmani can be bred successfully under suitable terrarium conditions. The clutch of two eggs is ready to hatch after about 60 days at 20 to 26°C (68 to 79°F). The freshly hatched young are 7.5 cm (3 in) long

PHOTO COURTESY OF FLUKER FARMS.

By feeding crickets a high-calcium diet, you will enhance their nutritional value to your anoles as well. Enriched cricket foods are available in many pet shops.

Male *Anolis garmani* presenting a beautiful throat-fan display. Note also the unscaled, regenerated section of the tail.

PHOTO BY R. D. BARTLETT.

PHOTO BY R. D. BARTLETT.

Anolis garmani can be found in Jamaica (where they are native) and in some parts of Florida (where they were introduced). In lowlands they live in tall trees and bamboo; in uplands they live in forests.

and display a brown ground color with crossbands. The parents can develop cannibalistic cravings for their young, therefore rearing should take place in separate small cages.

Anolis gingivinus

Description: *Anolis gingivinus* is today classified as an independent species, although formerly it was considered to be a subspecies of *Anolis bimaculatus*. *Anolis gingivinus* is more plainly colored: on the light brown to brownish gray ground color, the anoles exhibit washed-out dark brown markings over the entire dorsal side, including the flanks and the legs. On the side there is a dirty white lateral stripe that ends at the hind legs.

With a total length of 18 cm (7.2 in) in males and 14 cm (5.6 in) in females, *Anolis gingivinus* is one of the smaller species. Males have a yellow throat fan with light spots, and when excited they can erect a small fold of skin on the nape and the proximal part of the back.

Range and habitat: *Anolis gingivinus* is native to the Lesser Antilles. It inhabits the islands of St. Martin, St. Barthelemy, Anguilla, and Sombrero, where it is said to occur in trees and near the coast on rocks sprayed by sea water. As a follower of civilization, it is found in the vicinity of human habitation.

Care in the terrarium: *Anolis gingivinus* is kept in medium-sized terraria with a height of at least 80 cm (32 in). The anoles need temperatures above 30°C (86°F), as in the vicinity of an incandescent bulb or spotlight, but it can cool to about 23°C (73°F) at night. Through daily spraying we achieve the required relative humidity of 70 to 80 percent. Because of the small size of *Anolis gingivinus*, we should offer small food insects (meadow plankton, *Drosophila*).

Anolis griseus

Description: With a total length of up to 35 cm (14 in), *Anolis griseus* is a very large anole. The lizards, apart from the short and compressed head, have a slender build and conspicuously long legs. The ground color is a grayish brown and can take on slightly olive shades. Characteristic of *Anolis griseus* is a light lateral stripe set off by small dark spots, as well as a short white stripe above the front legs. In the nape region the anoles exhibit small, irregularly arranged black specks.

Anolis griseus.

ARTWORK BY JOHN R. QUINN.

PHOTO BY RALF HESELHAUS.

Anolis gingivinus is one of the more plainly colored Caribbean anole species.

The throat and the neck region are pale yellowish. Males and females have a gray throat fan with yellow or orange blotches and shadings. Females are similar in coloration to the males but are easy to distinguish from them by their smaller size.

Range and habitat: *Anolis griseus* is distributed throughout St. Vincent, but it is particularly abundant in the Kingstown Botanical Garden and in the Mesopotamian Valley. The anoles are found in large and tall trees, where they prefer to stay in the treetops.

Care in the terrarium: *Anolis griseus* needs a very large terrarium, in which the anoles are kept in pairs. The terrarium should be sparsely decorated with a few branches and plants, but nevertheless should offer

PHOTO COURTESY OF ENERGY SAVERS.

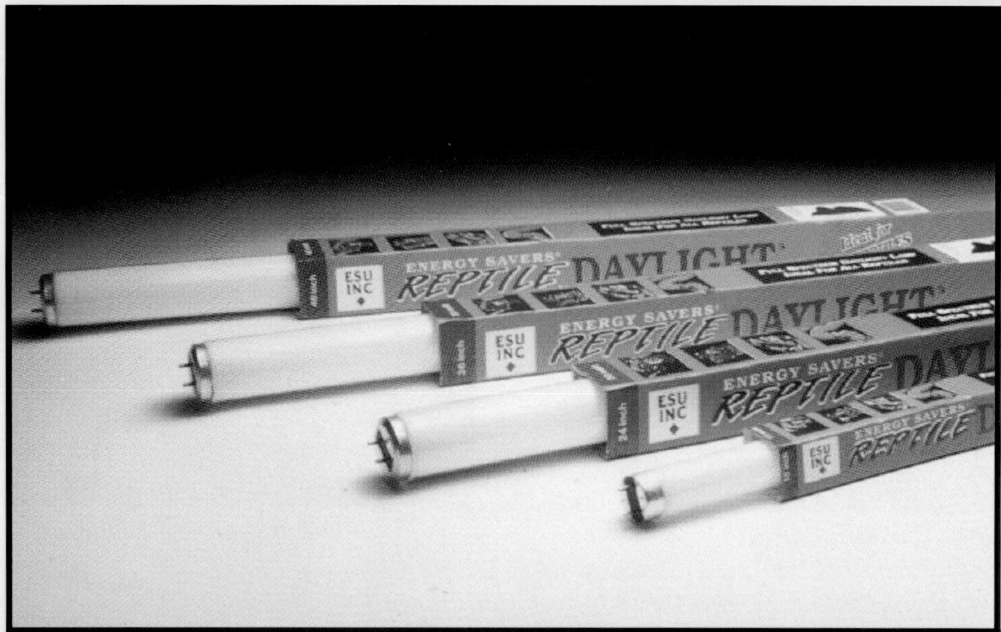

Providing Caribbean anoles with the correct photoperiod (day/night cycle) is very important. Photoperiod often is a factor in determining a herp's behavior. Bulbs designed specifically for the keeping of reptiles and amphibians now are available at many pet shops.

sufficient hiding places. *Anolis griseus* inhabits the same living space as *Anolis trinitatis* and should be treated in the same way. Because *Anolis griseus* also manages fairly large food animals, it should not be kept with smaller lizards.

Breeding is easy. About 45 days after copulation the female lays two eggs in the substrate. After about 50 days the almost 8-cm (3.2-in) young hatch. They are easy to rear.

Anolis luciae

Description: With a total length of 23 cm (9.2 in), *Anolis luciae* is a medium-sized anole. The anoles are able to undergo an extensive change in color, so that a yellow, brownish gray, black, or dark green ground color may be present. The legs often exhibit a slight blue tint. The throat fan is grayish white with a narrow black margin and occasionally exhibits small black spots. Females are predominantly light brown and exhibit a pattern of dark spots or stripes.

Range and habitat: *Anolis luciae* inhabits extensive parts of St. Lucia, though largely absent from the south-central area. Particularly popular living spaces are in the vicinity of villages and in hotel gardens. The population density is often so high that the anoles occur in virtually every tree and bush. A large tree is home, as a rule, to one adult male, several juvenile males, and ten or more females. The biotopes are relatively arid and windy.

Care in the terrarium: A well-lit and sparsely furnished terrarium is a prerequisite for keeping *Anolis luciae.* The decorations should consist of thick branches and pieces of bark adorned by robust plants. *Anolis luciae* needs daytime temperatures of

ARTWORK BY JOHN R. QUINN.

Anolis luciae.

about 30°C (86°F) and nocturnal cooling to 25°C (77°F). The humidity should be maintained at between 60 and 80 percent. For a breeding group, a sex ratio of one male to three females is recommended. With good and ample feeding—*Anolis luciae* is especially fond of small insects such as *Drosophila*—this species is not hard to breed.

Anolis marmoratus group

Between Dominica and Montserrat, in the northern section of the Lesser Antilles, lies the Guadeloupe Archipelago belonging to France. A total of 11 subspecies of the *Anolis marmoratus* group inhabits the island of Guadeloupe as well as various smaller surrounding islands. Of these, the larger islands of Iles des Saintes, Marie-Galante, and La Desirade are of special interest. The main island of Guadeloupe is divided into two clearly separated regions. Basse-Terre, the western part of the island, has a strongly mountainous character with active volcanoes that point to the geologic origin of the archipelago. At an altitude of 1467 meters, Basse-Terre towers over the low-lying eastern section of Grande-Terre. In regard to climate, as well, the wet Basse-Terre and the arid Grande-Terre represent two distinct zones. The two parts of the island, connected only by a narrow land bridge, each have a surface area of 40 X 25 kilometers. Six of the 11 *Anolis marmoratus* forms live here on the main island.

Anolis marmoratus alliaceus

Description: These very attractively marked anoles at times display clear differences in color distribution. The ground color consists of a rich and dark green, with no trace of the blue suffusion in the tail region often present in the other *Anolis marmoratus* forms. On the head region, which is more of a brownish green, there are blackish blue spots that can extend over the back to the tail region. The dark spots, all of which have cream-yellow margins, are so prevalent in some specimens that they can run into one

The range of 11 subspecies of *Anolis marmoratus*, plus the closely related *A. ferreus*.

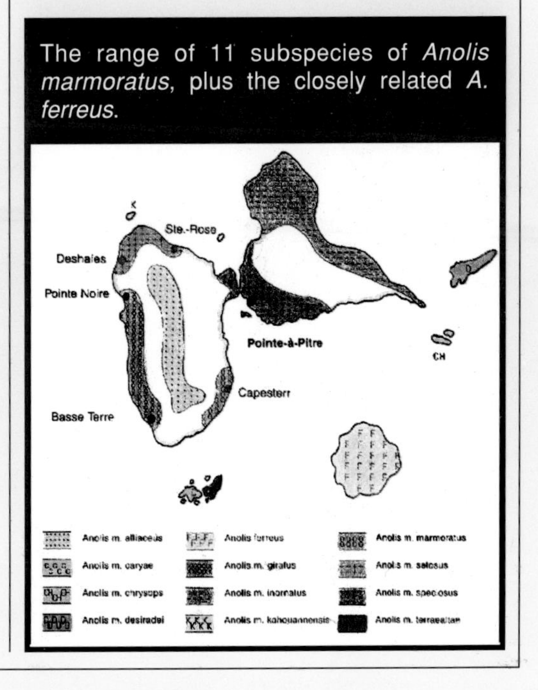

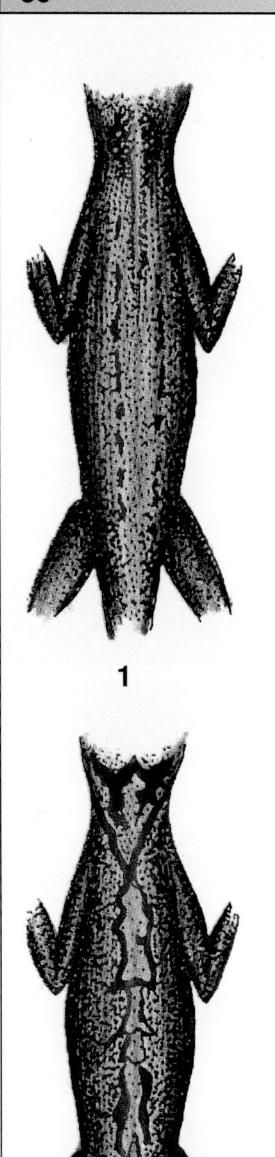

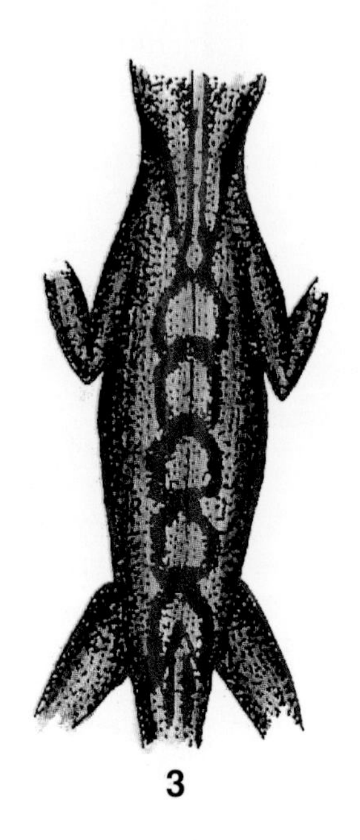

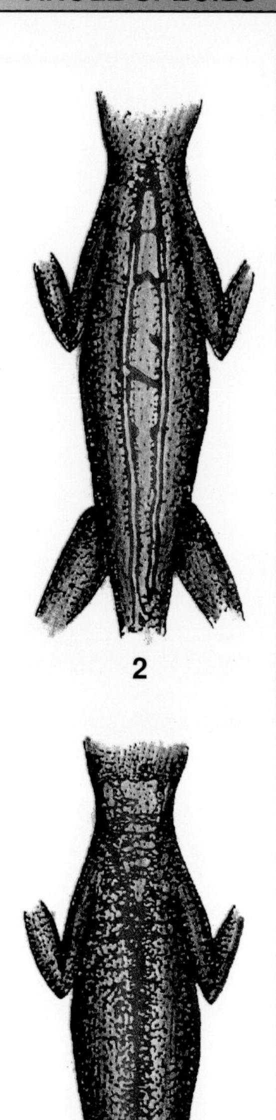

Dorsal markings of various *Anolis marmoratus* females.
1) *Anolis m. girafus*
2) *Anolis m. chrysops*
3) *Anolis m. caryae*
4) *Anolis m. alliaceus*
5) *Anolis m. desiradei*

another in the head, nape, and back regions, so that the anoles appear uniformly black in those areas. As a rule, the dark spots are distributed from the head region to the middle of the body, but some specimens exhibit only a spot in the shoulder region. *Anolis marmoratus alliaceus* has a dark orange throat fan with yellow-green shading. Also characteristic are two dark brown stripes extending through the eye and along the lower jaw. The ability to change color is limited to a darkening of the ground color, so that the cream-yellow margins of the speckling stand out even more strongly in contrast. The males are fully grown at a body length of 7.5 cm (3 in); the anoles reach a total length of more than 22 cm (8.8 in). Females are marked with pale green and exhibit distinct dorsal markings. The sides, with their small light and dark spots, stand out from the dorsal markings.

Range and habitat: This form lives exclusively in the central highlands, the rainforests of Basse-Terre, at altitudes between 600 and 900 m. *Anolis marmoratus alliaceus* prefers the tops of the up to 40-meter-high, epiphyte-covered trees. As a result of the relatively limited food supply in this region, the population density is rather low. In contrast to this are the areas where the rainforest borders small settlements. Next to cattle pastures, which are fenced in by thickly planted yucca trunks, *Anolis marmoratus alliaceus* finds optimal living conditions. The dense growth of plants offers excellent places to hide and, as a result of the cattle manure drawing insects, favorable feeding places. The anoles that occur in these places in great numbers look correspondingly robust and well-nourished.

Care in the terrarium: As a purely upland form, this anole naturally has very specific requirements. The temperatures in the terrarium should be between 22 and 25°C (72 and 77°F) during the day, cooling slightly at night to about 18 to 20°C (65 to 68°F). *Anolis marmoratus alliaceus* tolerates high humidity for fairly long periods of time, but dry sections always should be present in the terrarium.

Corresponding to the luxuriant plant growth in the rainforest, the terrarium should be thickly planted. *Anolis marmoratus alliaceus* soon loses its shyness in captivity and, despite numerous hiding places, will often show itself.

Anolis marmoratus caryae

Description: The ground color of this anole is a pale green, and the head, nape, and adjoining back are covered with a fine speckling, so that these areas take on a light gray shade. The belly is yellowish green. The whitish coloration of the chin and the throat is continued in a lateral stripe of the same color. The neck region is cream-colored and the throat fan is a rich yellow. The maximum body length is 7.5 cm (3 in), and the total length is up to 22 cm (8.8 in). Females are pale gray-green. They display pronounced dorsal markings and have a lateral stripe that is bordered on both sides by a fine dark speckling.

Range and habitat: *Anolis marmoratus caryae* lives on the islands Iles des Saintes and Terre-de-Bas, where the anoles also inhabit, in addition to the arid regions, somewhat wetter regions. The idyllic islands also have tall-growing stands of trees as well as rather extensive palm groves that form the habitat of this subspecies.

Care in the terrarium: *Anolis marmoratus caryae* should not be kept in a damp terrarium. Although the anoles tolerate high humidity for short periods of time, they cannot do so in the long run. Therefore, *Anolis marmoratus caryae,* like all other subspecies of *Anolis marmoratus* of the satellite islands around Guadeloupe, should be kept in a well-ventilated terrarium that can, however, safely be sprayed once a day. Large ventilation openings should provide for rapid drying of the dampness and therefore for an optimal microclimate in the terrarium.

Anolis marmoratus chrysops

Description: *Anolis marmoratus chrysops* is a dark gray-green on the body. The sides display areas of yellow, and the belly is bright yellow. The anoles exhibit speckling in the nape and dorsal region. Like *Anolis marmoratus desiradei, Anolis marmoratus chrysops* also displays red coloration in the neck region, which also has a small yellow component, and a brownish gray head. On the other hand, the blue-gray chin region points to a kinship with *Anolis ferreus* from the neighboring island of Marie-Galante.

Bulbs designed specifically for the keeping of reptiles and amphibians now are available at many pet shops. Some provide not only light but a measure of heat as well.

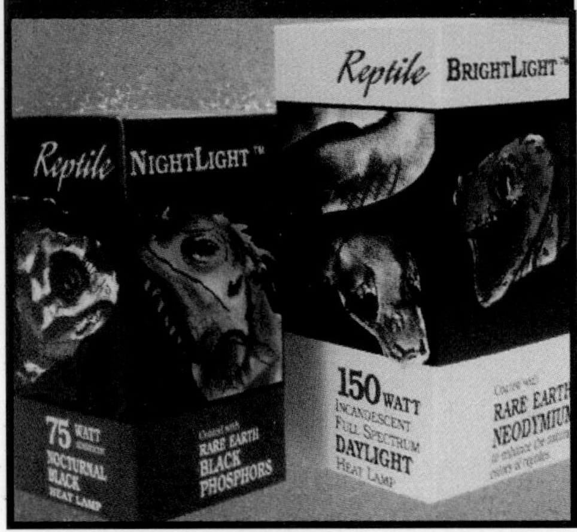

PHOTO COURTESY OF ENERGY SAVERS.

The males of *Anolis marmoratus chrysops* have a rich yellow throat fan with a few pale yellow patches and reach a total length of 20 cm (8 in) or more. The ground color of the female is gray and is broken up only by faint suggestions of dark longitudinal stripes on the back and a short stripe on the side. The belly is yellow.

Range and habitat: *Anolis marmoratus chrysops* lives on the islands of Iles de Petite Terre, Terre-de-Haut, and Terre-de-Bas southeast of Guadeloupe. *Anolis marmoratus chrysops* is also subject to special conditions in its natural habitat: the small and flat islands are strongly influenced by the trade winds. Low-growing vegetation and an arid climate are typical.

Care in the terrarium: *Anolis marmoratus chrysops* under no circumstances will tolerate high humidity in conjunction with poor air circulation. With this subspecies we need a terrarium with large ventilation surfaces to prevent the stagnant air that is deadly to the anoles. The temperatures should be about 28°C (82°F) during the day and 22°C (72°F) at night. The humidity can briefly (after spraying) attain values above 80 percent.

Anolis marmoratus desiradei

Description: The ground color of the male is a pale green with a yellowish suffusion on the sides. Characteristic are numerous dark grayish brown patches in the dorsal region and on the legs. Because of the absence of markings on the flanks, a light stripe is suggested. The ventral side is uniformly yellow. The head region is a rich brown, which turns to an intense rusty red in the neck region and stands out in clear contrast to the greenish ground color. The throat fan is

ARTWORK BY JOHN R. QUINN.

Anolis marmoratus desiradei.

yellow. A color change in this subspecies leads to a darkening of the various shades of brown, which then makes the dorsal markings stand out more strongly. This subspecies may reach 21 cm (8.4 in) in total length. Females have a gray head coloration that merges into the greenish body coloration. Dorsal markings are absent. The belly is yellow.

Range and habitat: *Anolis marmoratus desiradei* lives exclusively on the island of La Desirade east of Guadeloupe. The anoles have no centers of distribution, but rather are distributed uniformly over the island. The habitat is similar to that of *Anolis marmoratus inornatus* on Grande-Terre.

Care in the terrarium: *Anolis marmoratus desiradei* is a suitable terrarium charge and makes no special demands. In the same way as with the other *Anolis marmoratus* subspecies from the satellite islands of Guadeloupe, good ventilation of the terrarium is required.

Anolis (marmoratus) ferreus

Description: *Anolis ferreus,* formerly known as *Anolis marmoratus ferreus,* is a particularly robust and large anole. Fully grown males reach a total length of 35 cm (14 in). Females, on the other hand, are distinctly smaller. Male *Anolis ferreus* have a predominantly gray-brown ground color. The sides of the anoles, in particular, are often yellowish or ocher. The belly is uniformly pale yellow. The head region is blue-gray. The throat fan is predominantly pale yellow with dark yellow patches. Some individuals display dark speckles extending from the head to the dorsal region. The tail gradually changes from the brownish gray of the dorsal markings into a brownish, washed-out blue-green. Particularly striking in full-grown *Anolis ferreus* males is the distinct, high crest on the proximal part of the tail, through which run several dark brown longitudinal stripes. Females are in general brownish green, but can also be gray-brown to ocher, and exhibit a short flank stripe in the shoulder region. No throat fan is present in the females.

Great differences in form and coloration can occur in both sexes. For example, the form and size of the male crest vary greatly from individual to individual. Moreover, the coloration of the throat fan is not always uniformly yellow. Some individuals feature brown patches, and others have a blue-gray throat fan. There even are individuals with a bicolored throat fan, one half gray, the other yellow. Females vary especially in their ground color, which ranges from light and dark shades of brown to blue tones.

Range and habitat: About 20 kilometers south of Guadeloupe lies the small, almost circular island of Marie-Galante. It is an island with a

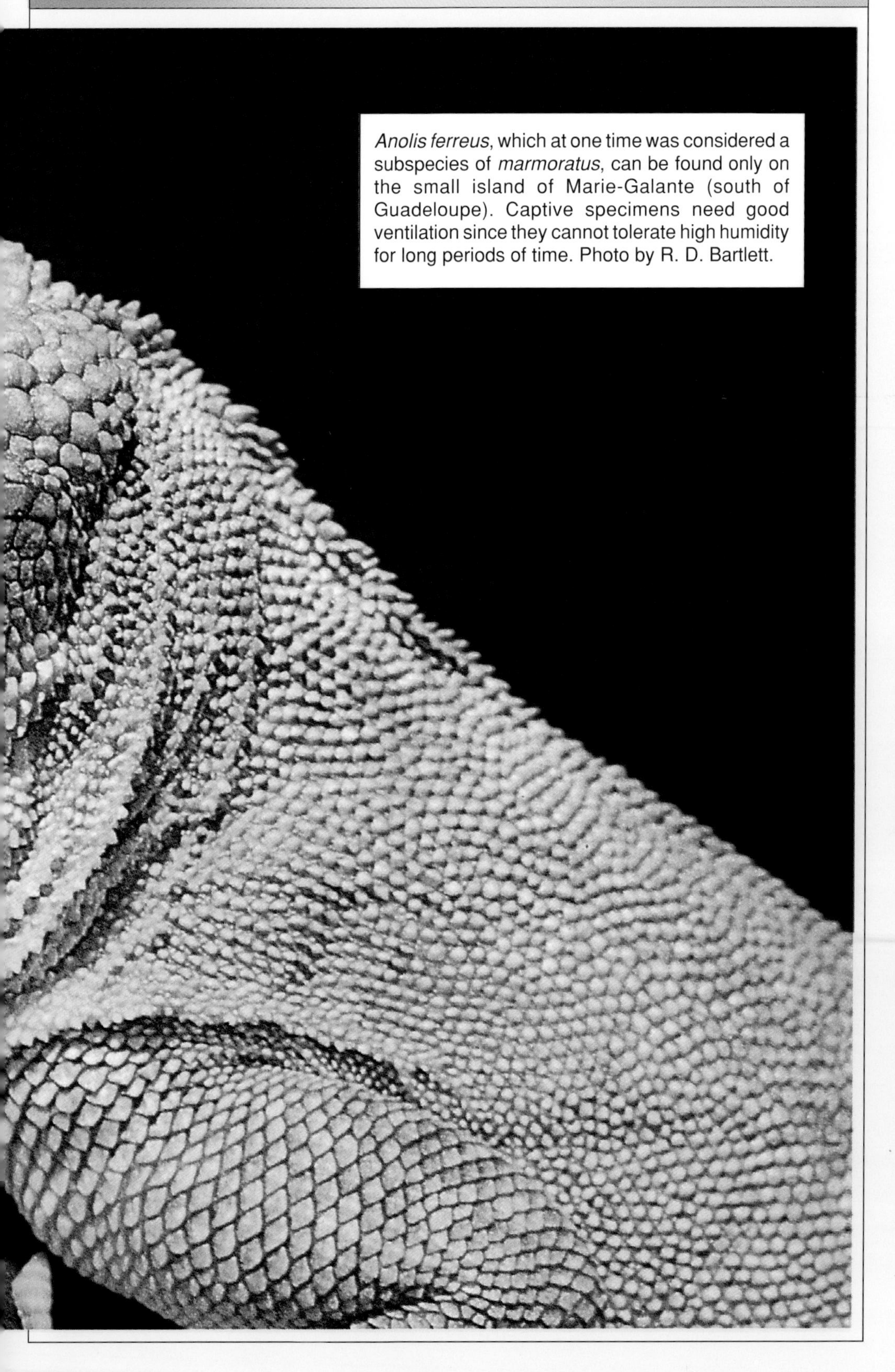

Anolis ferreus, which at one time was considered a subspecies of *marmoratus*, can be found only on the small island of Marie-Galante (south of Guadeloupe). Captive specimens need good ventilation since they cannot tolerate high humidity for long periods of time. Photo by R. D. Bartlett.

very arid climate. Extensive areas are covered by low bushes and thorny scrub, and only isolated wooded regions are present. *Anolis ferreus* clearly favors regions with the biggest, tallest trees. Furthermore, the stand of trees must not be too thick. We also find *Anolis ferreus* in hedgerows and stands of younger trees, although here they are mostly juveniles. *Anolis ferreus* has no particular center of distribution on Marie-Galante, and the anoles range over the whole island.

Care in the terrarium: On account of its size, *Anolis ferreus* needs a roomy terrarium that, if possible, should not have dimensions smaller than 80 X 80 cm (32 X 32 in). Ventilation is of decisive importance for the successful care and breeding of this species. The terrarium should be constructed such that at least one side is completely screened. To achieve air circulation, at least one other ventilation opening should be installed in the opposite side. *Anolis ferreus* needs especially good ventilation and in no case tolerates high humidity for very long.

Anolis ferreus prefers terraria that are not too thickly planted and decorated. Because the anoles are very active, only robust plants are suitable; they can easily be used to create suitable hiding places, which all anoles require. The ideal group size is one male and two females.

PHOTO BY R. D. BARTLETT.

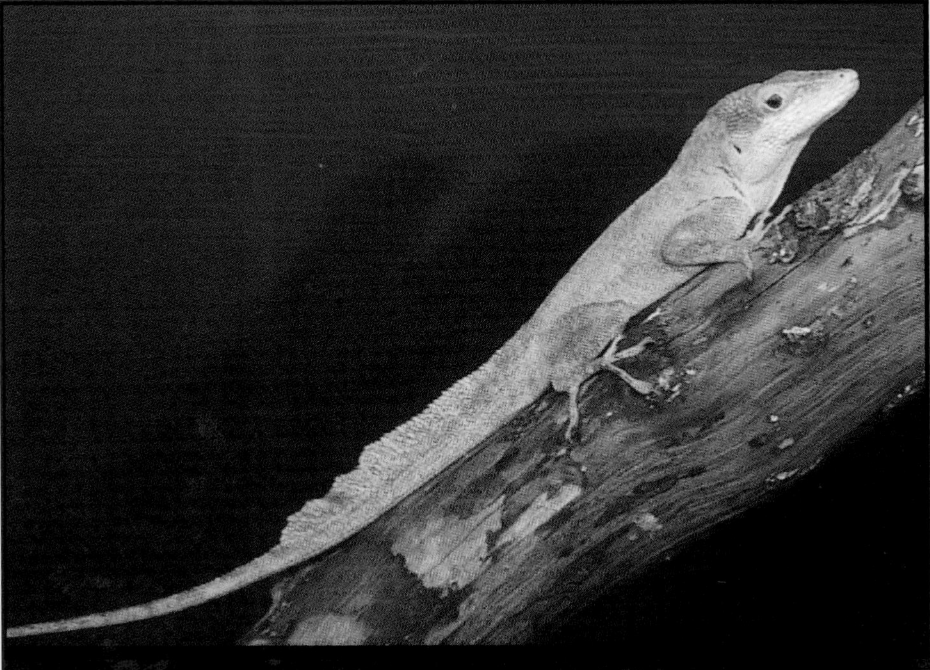

Use only robust plants when decorating the enclosure of *Anolis ferreus*. These anoles are very active and will destroy weaker plants.

Anolis marmoratus girafus

Description: Adult males, which reach a size of 23 cm (9.2 in), have a brown head and nape coloration that merges through a beige color into the blue-green of the body. In the area of the base of the tail the anoles can be bluish in color. On this ground color they exhibit variable markings that usually consist of brown or beige, sometimes extensive, speckling. The belly is dirty white and the throat fan is yellow with white shading. The females vary in coloration from rich shades of brown to gray. The dorsal markings consist of small, dark spots arranged in lines that extend in two broken rows from the nape to the tail region.

Range and habitat: *Anolis marmoratus girafus* inhabits the western coast of Guadeloupe from the town of Basse-Terre north to about Pointe Noire. On account of its location on the leeward side of the mountains, this is the most arid region in all of Basse-Terre. As a follower of civilization, the anole lives here preferentially in villages on house walls, lamp posts, fences, and other structures. The anoles also are present along the coastal road in rural regions. They inhabit the large trees on the side of the road, although the population density is relatively low.

Care in the terrarium: *Anolis marmoratus girafus* is not hard to keep. As with all *Anolis marmoratus* varieties, the terrarium must not be too small and should be well ventilated. Dimensions of 80 X 80 cm (32 X 32 in) are adequate. A special aspect of keeping this subspecies is based on the aridity of the natural habitat. Naturally the terrarium climate also should be correspondingly arid.

PHOTO BY R. D. BARTLETT.

Anolis ferreus seems to have no particular center of distribution on its home island of Marie-Galante, but specimens seem to favor areas with tall trees. Juvenile specimens, appropriately enough, often are found in the stands of smaller trees.

Anolis marmoratus inornatus

Description: The ground color of full-grown males is a pale gray-green that merges into a rich brown toward the head. In some individuals the brown of the head coloration is repeated in suggested dorsal markings. A flank stripe, lighter gray-green than the dorsal coloration, runs from the shoulders to the hind legs. The belly is a pale yellow-green. Some individuals exhibit a slight blue suffusion in the tail region. The throat fan is intense yellow and can exhibit white patches. *Anolis marmoratus inornatus* can reach a body length of 7.5 cm (3 in); the total length is 22 cm (8.8 in). Apart from the brown head color, the females are more of an ashy gray. A variably expressed dorsal pattern is always apparent, and at least the suggestion of a flank stripe is present.

Range and habitat: This form inhabits northern and northeastern Grande-Terre. The range extends along the northern coast to the Pointe des Chateaux. Individuals from Pointe des Chateaux represent a transitional form to *Anolis marmoratus desiradei.* The habitat of *Anolis marmoratus inornatus* is similar to that of *Anolis marmoratus speciosus* but is even more arid and is characterized by bushy vegetation.

Care in the terrarium: The terrarium should contain the same arid conditions as those of Grande-Terre. Good ventilation is a prerequisite for the successful keeping and breeding of this subspecies of *Anolis marmoratus.*

Anolis marmoratus kahouannensis

Description: *Anolis marmoratus kahouannensis* appears to be almost of one color. Only on the chin and occasionally other parts of the head does a slight blue shading occur. The rest of the body is uniformly colored in rich green. The throat fan is yellow, the belly pale yellow. The change in color by this variety, which with a total length of nearly 24 cm (9.6 in) is relatively large, consists only of a darkening of the green. Like the males, full-grown females exhibit absolutely no markings and are plain green. Only juveniles may exhibit dorsal markings.

Range and habitat: *Anolis marmoratus kahouannensis* lives exclusively on the small islands of Ilet-a'-Kahouanne and Tete-a-Anglais, located off the northern part of Basse-Terre. The biotope resembles closely that of *Anolis marmoratus setosus.*

Care in the terrarium: *Anolis marmoratus kahouannensis* presents no special problems in keeping and breeding. The terrarium requirements are similar to those of *Anolis marmoratus setosus.*

Anolis marmoratus marmoratus

Description: With a total length of 24 cm (9.6 in), *Anolis marmoratus marmoratus* is a rather robust anole. It is often claimed that *Anolis marmoratus marmoratus* is the most beautiful Caribbean anole, and it is true that males display a particularly rich and contrasting coloration. The ground color is a light but intense green that merges into turquoise to blue toward the base of the tail. The base coloration of the head region is blue-gray and is covered by numerous bright orange-red blotches. The large throat fan is orange-yellow with small yellow blotches. At night this anole is very dark in color. The general body coloration is then a dark chocolate brown, and only the darkened orange spots still suggest the gorgeous colors that *Anolis marmoratus marmoratus* can reveal during the day. The females of this subspecies are grassy green on the sides and are gray-brown on the head, the back, and in the tail region. The gradations in color flow into one another, so that no contrasting striping or specific dorsal markings are present.

are locations exposed to the wind and open plant growth.

Care in the terrarium: *Anolis marmoratus marmoratus* can be a good terrarium charge if several significant points are considered. For example, the terrarium, in keeping with the body size of the anoles,

PHOTO BY MATTHIAS SCHMIDT.

Anolis marmoratus marmoratus is one of the more attractive *marmoratus* races. It lives along the eastern coast of Basse-Terre, where it often is found in banana plantations.

should be large. Planting is unsuitable if it consists of small, delicate plants. Some branches and a few yuccas must suffice here. *Anolis marmoratus marmoratus* needs its hiding places, but the terrarium must be open and sparsely decorated.

The most important point for the successful care of these anoles is adequate ventilation. Only in this way can the high humidity otherwise customary in tropical terraria be decreased. Stagnant air in association with high humidity means a fatal end for *Anolis marmoratus marmoratus* after a short stay in the terrarium!

Range and habitat: *Anolis marmoratus marmoratus* is found along the eastern coast of Basse-Terre from Capesterre to Bananier. The entire coastal strip is characterized by large banana plantations that form part of the habitat of the anole. Here live smaller males and many juveniles. Other popular sites are fence posts in pastures when the fence is bordered by light shrubbery. The especially beautifully colored specimens favor the big trees that line the avenues near towns. Of significance for all *Anolis marmoratus marmoratus* biotopes

Anolis marmoratus setosus

Description: Full-grown males exhibit a green or blue-green dorsal coloration. In the head region they can even look blue. A few individuals exhibit extensive arrangements of small yellow spots in the nape region. The belly is pale green. The throat fan is intense yellow with greenish or gray-green patches.

Anolis marmoratus setosus is able to undergo an extensive change in color to dark shades, so the anoles can turn almost black, particularly on the back. During the color change, dark crossbands become particularly prominent on the back. The virtually black eye stripe is also conspicuous during the change in color. The maximum body length of these relatively small anoles is only 6.6 cm (2.6 in); the total length is 19 cm (7.6 in).

The females have weak pied markings. Shades of green and brown predominate, although various gradations of gray and white also are present. Females do not exhibit uniform dorsal markings; instead, rows of dark spots on the sides suggest a striped pattern on the otherwise dirty white ground color.

Range and habitat: *Anolis marmoratus setosus* lives as a pure form exclusively in northwestern Basse-Terre, approximately between Ste.-Rose and Deshaies. The range extends into the ranges of *Anolis marmoratus girafus, Anolis marmoratus alliaceus,* and *Anolis marmoratus speciosus,* and a considerable mixing of the varieties occurs. Accordingly, the safe identification of the subspecies in the transitional zone is quite difficult.

Anolis marmoratus setosus lives preferentially on the edges of towns in the vicinity of buildings as well as in open forests along the coast. The biotope is not as arid as that on Grande-Terre, and the vegetation consists largely of extensive wooded areas. Generally speaking, the population density is not very high.

Care in the terrarium: Like *Anolis marmoratus speciosus, Anolis marmoratus setosus* also is a highly adaptable anole. The anoles can be kept and bred in pairs even in fairly small terraria. The terrarium should be sparsely decorated and furnished with not too many climbing branches.

Anolis marmoratus speciosus

Description: *Anolis marmoratus speciosus* is an especially beautiful

A female *Anolis marmoratus speciosus* exhibiting sunken flanks immediately following egglaying.

PHOTO BY MATTHIAS SCHMIDT.

anole. Males exhibit an extremely intense green that is not broken up by markings on the sides or back. The belly is yellowish, and in the head region the anoles often exhibit a sky-blue coloration. The blue to turquoise color can also be present on the hind legs and the tail. The anoles measure about 7 cm (2.8 in) from the head to the base of the tail; the total length is about 21 cm (8.4 in). The sulfur-yellow throat fan features small greenish patches. In contrast to *Anolis marmoratus setosus, Anolis marmoratus speciosus* never has a dark eye stripe. The change in color is limited mainly to a darkening of the green ground color; in rare cases a change to a gray color occurs. The olive-green females exhibit as dorsal markings a pattern of stripes or freckles.

Range and habitat: *Anolis marmoratus speciosus* occurs throughout southwestern Grande-Terre, on the narrow land bridge between Grande-Terre and Basse-Terre, as well as on the island of Ilet du Cochons. Within this region there are numerous centers of distribution. The flat part of the island is used mainly for growing sugar cane and can in general be described as arid. Even in the rainy season there is rarely any persistent moisture. The often sparse vegetation and the fresh ocean wind quickly dry up the water. The highest population densities are attained in parks or other green areas. Here the bright green males live in almost every tree, where they avidly court the females.

Care in the terrarium: *Anolis marmoratus speciosus* is an extremely hardy terrarium charge. This subspecies tolerates fairly large fluctuations in temperatures and humidity very well. *Anolis marmoratus speciosus* can be kept and bred under both high humidity and arid conditions.

The rivalry between individuals is not too intense, but several males can be kept together only in really large terraria. As a rule, the ratio of one male to two or three females is recommended in the terrarium.

PHOTO BY MATTHIAS SCHMIDT.

In typical posture, *Anolis marmoratus speciosus* "stands" in its territory.

Anolis marmoratus terraealtae

Description: This anole is grayish brown or greenish brown in color. As a rule, the head region is dark brown; on the side of the head and in the nape region several fairly large yellow spots are present. The coloration in the tail region tends to blue-green. *Anolis marmoratus terraealtae* attains a maximum length of nearly 23 cm (9.2 in). The male's throat fan is orange-yellow with patches of yellow. Females are pale grayish brown in color and exhibit distinct striped markings on the back. The anoles have a short stripe on the sides.

Range and habitat: *Anolis marmoratus terraealtae* lives in the low arid regions of the islands lying to the south of Basse-Terre: Ilet-a-Cabrit, the eastern part of Terre-de-Haut, Grande Ilet, and Iles des Saintes. These regions border on the high and wet western end of Terre-de-Haut and are located east of the likewise high and wet island of Terre-de-Bas. *Anolis marmoratus terraealtae* exhibits the greatest population density on large trees in the vicinity of buildings.

Care in the terrarium: From the area of occurrence it can be seen that *Anolis marmoratus terraealtae* needs an arid climate. If the basic requirements for keeping in the terrarium are present, this species is fairly hardy.

Anolis oculatus group

Dominica, the northernmost of the Windward Islands, lies between the French islands of Martinique and Guadeloupe and is located about 50 km from each. The 50-km-long and 22-km-wide island, which is of purely volcanic origin, features high mountains covered by dense rainforest, hundreds of waterfalls, crater lakes, and sulfur springs. On the island, still relatively undisturbed today, four subspecies of *Anolis oculatus* have evolved: *Anolis oculatus oculatus, Anolis oculatus cabritensis, Anolis oculatus montanus,* and *Anolis oculatus winstoni.*

Anolis oculatus cabritensis

Description: The up to 23-cm (9.2-in) males are predominantly gray or pale yellow-brown. The belly is yellowish or peach. As a rule, *Anolis oculatus cabritensis* displays two or three conspicuous black spots on the sides. Some populations can also exhibit extensive markings on the sides that are broken up only by a pattern of fine, light spots or lines. The head and nape region are distinguished by numerous small white spots and stripes. In many individuals the head region is darker, occasionally reddish in color. The blue eyes characteristic of *Anolis oculatus cabritensis* provide an interesting contrast to this coloration. The throat fan, as with *Anolis oculatus oculatus,* is yellow or orange.

Individuals from the region around the village of Picard have a somewhat darker and more yellowish ground color than do the previously described variants from the Cabrits. Anoles from the region around Pointe Ronde are conspicuously yellow, particularly on the sides. Populations from the region of Grand Savanna, on the other hand, are the lightest specimens. The ground color is a pale gray yellow; the belly is yellow. Although the somewhat darker head region still stands in contrast to the remaining coloration, the red component has disappeared almost entirely.

The coloration of the females consists of a gray ground color with yellowish sides. Both the females and juveniles of this subspecies exhibit a considerably more pronounced pattern of spots than do the other *Anolis oculatus* subspecies. The dorsal and lateral stripes are also more pronounced.

Range and habitat: *Anolis oculatus cabritensis* lives along the western coast of Dominica from the Cabrits to Grand Savanna. This is one of the most arid regions of

Dominica, although clear climatic differences do occur in this range, particularly in regard to precipitation and the vegetation. For example, the area around Grand Savanna is relatively arid and the region farther to the north around Picard is rather damp.

Care in the terrarium: *Anolis oculatus cabritensis* is a very interesting as well as attractively colored terrarium charge. To ensure proper care, it is essential to know the provenance of this anole. Individuals from the northern part of the range are less sensitive to high humidity than are those from southern populations, for which high humidity is fatal in the long run.

A good breeding group is achieved with a ratio of one male to two females. The robust anoles, which need lots of food and do not pass up even fairly large morsels, can become almost finger-tame after a brief acclimation period and take food from the keeper's hand. With good care in a roomy terrarium, the regular production of offspring can be expected.

PHOTO BY MATTHIAS SCHMIDT.

An attractive male specimen of *Anolis oculatus cabritensis*. In order to keep this animal healthy in captivity, a keeper should know the precise origin of the specimen. Specimens from the southern part of the *cabritensis* range are more sensitive to humidity than those from the north.

Anolis oculatus montanus

Description: The ground color of this big anole varies from light to dark shades of green. The sides are not as intensely green as the back, but they can take on a metallic sheen. Numerous light, almost white, spots of variable size form distinct stripes. The first stripe usually consists of larger, more widely separated spots, while the remainder consist of small, closely spaced dots. In the area of the first row, several black spots can be visible. A particularly intense speckling is present on the head and nape. The throat fan is dark yellow with rusty areas. Characteristic of *Anolis oculatus montanus* is the well-developed tail crest. With a total length of up to 28 cm (11.2 in), it is the largest *Anolis oculatus* subspecies. The females and juveniles have a green ground color. The sides, as with the male, have a metallic luster. The anoles have numerous small spots and exhibit dorsal and lateral stripes. The female's throat fan is dark red.

Range and habitat: The anoles inhabit the entire central highlands of Dominica. *Anolis oculatus montanus* is thus an upland form and is found only at altitudes between 600 and 900 m. High humidity prevails in this rainforest region. The biotope features numerous watercourses and waterfalls, where *Anolis oculatus montanus* prefers to stay near water on large rock formations.

Care in the terrarium: *Anolis oculatus montanus* needs lots of space. The terrarium should have as large a ground space as possible and therefore in most cases should be longer than high. To reach the required humidity, a large aquatic section should be installed in such a way that it is well-hidden by suitable plantings and equipped with overhanging branches.

The most desirable temperatures are between about 20 and 25°C (68 and 77°F). *Anolis oculatus montanus* should not be kept in groups of more than one male and three females.

Anolis oculatus oculatus

Description: The full-grown males of *Anolis oculatus oculatus* vary greatly in color. All that they share is the light olive, more yellow-brown, ground color that is an ideal adaptation to their surroundings. The belly is white to dirty yellow. Many individuals display a faint yellowish crossbanding on the body. Just behind the front legs, one, two, or occasionally even three black spots can be present on the side. Some individuals exhibit no black pigmentation at all. The throat is yellowish orange. In contrast to the other subspecies, *Anolis oculatus oculatus* has only traces of a crest on the nape and tail, and with a length of 21 cm (8.4 in) it is the smallest subspecies. The females are uniformly olive to yellow-brown. As a rule they have faint dorsal stripes.

Range and habitat: The arid southern and southwestern sides of Dominica are the range of *Anolis oculatus oculatus*. Limited by the Morne Paix Bouche in the southeast, which extends directly to the coast and forms both a geographic and climatic barrier, the anole's range extends to the west about as far as the Layou Valley. Because of the windward location of the range, the habitat of *Anolis oculatus oculatus* has more of a steppe-like character. Arid regions with intense sunshine and only scrubby vegetation characterize this zone in southern Dominica.

Care in the terrarium: When the climatic conditions are considered, *Anolis oculatus oculatus* is easy to keep and breed in the terrarium.

Anolis oculatus winstoni

Description: *Anolis oculatus winstoni* probably is the most beautifully colored and richly contrasting subspecies. The rusty brown ground color, which is broken up by numerous small to medium-sized marks over the whole body, merges into peach to yellow shades on the sides. The white marks, which are densest in the head and nape region, can be separated by fine black lines. The skin color around the blue eyes varies from white to sky-blue. The large throat fan is rich orange-yellow. The males have a well-developed nape and tail crest and reach lengths of 22 cm (8.8 in). The relatively large and robust females are likewise rusty brown, and only occasionally is a slight tendency to olive shades present. The white marks arranged in lines are less pronounced than with the males.

Range and habitat: This variety has by far the most extensive range

BOTH PHOTOS BY MATTHIAS SCHMIDT.

Above: A full-grown male *Anolis oculatus winstoni.* **Below:** The robust females of *winstoni* are rather plainly colored.

on Dominica. The anoles inhabit the entire eastern coast from Penville in the north to the natural barrier of the Morne Paix Bouche in the south. *Anolis oculatus winstoni* inhabits the coastal region and is also rather abundant inland in flat regions. The entire eastern coast is very wet because of the constant northeast wind. The vegetation is accordingly lush and rich in species. *Anolis oculatus winstoni* favors mainly dark biotopes near water, where the temperature during the day is about 25°C (77°F). Large trees, which have a bizarre appearance due to aerial roots, as well as rocks near water are favored territorial sites.

Care in the terrarium: *Anolis oculatus winstoni* is easy to keep and breed. The beautiful anoles are neither fussy in regard to food nor are they sensitive to temperature fluctuations. They prefer a high humidity, which can be nearly 100 percent at night but should decrease to 50-60 percent during the day. This means that *Anolis oculatus winstoni* also requires a relatively well-ventilated terrarium in which dry places must be present. This subspecies is prone to skin diseases when subjected to damp, stuffy conditions; such diseases can be fatal without appropriate treatment.

The not-too-small terrarium can be decoratively furnished and planted. Because there rarely is a chance of finding an egg clutch in such heavy planting, the youngsters, which will certainly turn up in the terrarium eventually, should be transferred as quickly as possible to suitable rearing terraria. A good breeding group consists of one male and two females.

Anolis porcatus

Description: *Anolis porcatus* resembles the Green Anole, *Anolis carolinensis,* because it is likewise bright green in color and has a comparable build. This species also has the ability to change color rapidly from green to dark brown. At a total length of 25 cm (10 in), *Anolis porcatus* grows larger than *Anolis carolinensis,* and it is altogether more robust. Males have a yellow to pink throat fan; only the suggestion of a fan is present in females. In display coloration the male shows patches of blue on the nape and around the eyes. The belly is white in both sexes.

Range and habitat: *Anolis*

Anolis porcatus has the ability to quickly change color from green to brown. Specimens often are seen scurrying around areas heavily populated by humans (hotels, parks, etc.).

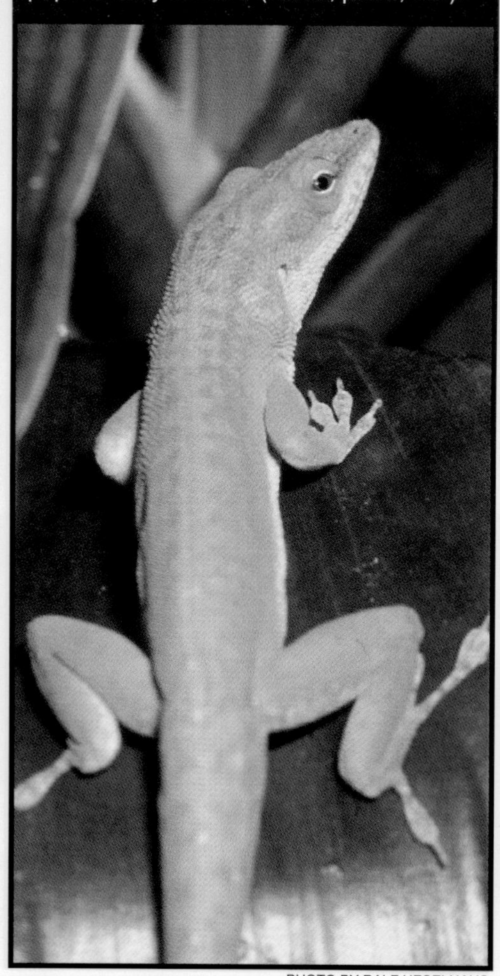

PHOTO BY RALF HESELHAUS.

porcatus is distributed over much of Cuba, both on the main island and the surrounding smaller islands. It lives in open forest and also in the immediate vicinity of humans in hotel gardens and parks. On the Isla de la Juventud we found the species in the garden of our hotel in palms, green shrubs, and also occasionally in the room of our bungalow.

Care in the terrarium: *Anolis porcatus* needs a medium-sized terrarium about 90 cm (36 in) high. The temperature should be 28°C (82°F) during the day, with a nightly cooling to 22°C (72°F). Because the anoles like to bask, a spotlight or an incandescent bulb in the terrarium cover (protected by screen) is recommended. The diet consists of the usual insect food, and *Anolis porcatus* should be fed only every other day.

Following hibernation in winter (two months at reduced temperatures and little food), the males soon come into breeding condition in the spring. Copulation occurs in the manner customary of anoles. A few weeks later the females lay two eggs in damp substrate. Maturation is best carried out in an incubator. The young hatch after about 60 days at 25°C (77°F) and are reared individually with tiny insects and the daily administration of a calcium preparation.

Anolis richardi

Description: With a length of more than 30 cm (12 in), *Anolis richardi* is a large, streamlined anole. The head and nape regions are light gray, and the body color is olive green. The long legs are a somewhat darker olive-green than the body. The tail is a light gray-green. Characteristic for *Anolis richardi* is a light lateral stripe that forms a strong contrast to the olive-green ground color of the back. Toward the belly, however, it merges almost without contrast into the only slightly darker ventral coloration. The throat is orange-yellow in both sexes and is equipped with gray-green patches and shadings. The sexes are easy to recognize by the clear size difference, females being only 15 cm (6 in) long.

Range and habitat: *Anolis richardi* lives on the islands of Grenada, Tobago, and the Grenadines, as well as the island of Bequia belonging to St. Vincent, from which a few individuals have also spread to St. Vincent. The habitat and the behavior of *Anolis richardi* are comparable to those of *Anolis griseus* on St. Vincent.

Care in the terrarium: *Anolis richardi* should be kept in roomy terraria and cared for as with *Anolis griseus*. Constantly damp areas should be avoided in the terrarium, and good ventilation is essential. *Anolis richardi* is an extremely adaptable anole that is easy to keep and can also be bred when the breeding stock is put together in pairs.

Anolis roquet group

In the middle of the Lesser Antilles lies Martinique, an island about 80 km long and 20 km wide. With Mt. Pelee, Martinique attains its highest elevation of 1464 meters. The geographic location of the island ensures a warm tropical climate with year-round temperatures of 24 to 28°C (75 to 82°F). Only during the months of April and May and between August and October does an often wet climate with frequent rainfall prevail. This also is the time of the feared West Indian hurricanes. During the remaining months it is rather dry on this Caribbean island, and only the ever-present trade winds provide a little cooling. Because of human development, only the high

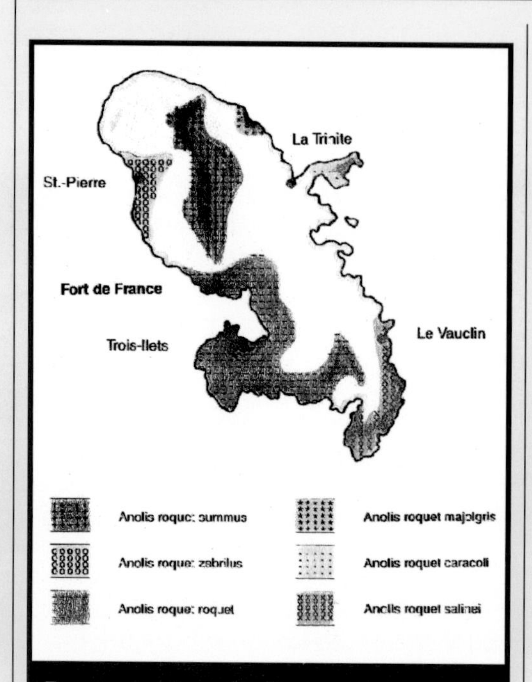

Range of various *Anolis roquet* races.

elevations are still covered by lush, undisturbed vegetation. In complete contrast to the deep green of the montane forests, the rest of Martinique reveals a totally different side. Sugar cane and banana plantations, pineapple fields, and extensive arid areas with scrub characterize the island. Only in a few places do coconut palms fringe the otherwise rather dreary coastal strip.

Anolis roquet is represented by six subspecies on Martinique. As a result of the enormous adaptability of the anoles to their environment, they have been able to evolve into pure followers of civilization. Accordingly, the extensive cultivation and alteration of the environment have not been detrimental to the anoles on Martinique. To the contrary, the highest population densities are found in the inhabited regions.

Anolis roquet caracoli

Description: *Anolis roquet caracoli* is a rather plain brownish gray. Characteristic of the subspecies are numerous dark brown markings on the head and body. The light gray coloration of the head region can extend well onto the throat fan, which is predominantly yellowish. Full-grown males reach a length of about 19 cm (7.6 in); females reach up to 14 cm (5.6 in). Females are a uniform gray with a small gray throat fan.

Range and habitat: This rather rare and wary subspecies is found only on Presqu'ile de la Caravelle in northeastern Martinique. *Anolis roquet caracoli* lives in the vicinity of

Anolis roquet caracoli is one of the rarer Caribbean anoles, being found only in northeastern Martinique. In captivity this animal will spend much of its time in hiding if its keeper provides too much light, thus a dim bulb is recommended.

PHOTO BY MATTHIAS SCHMIDT.

old ruins and in small forest openings. The anoles are not particularly partial to sun, but rather prefer shady places. The population is quite high in the small area of occurrence.

Care in the terrarium: *Anolis roquet caracoli* prefers relatively dim light, temperatures of about 26°C (79°F), and a humidity between 60 and 80 percent. The terrarium should not be too thickly planted; instead, the anoles need a robust network of branches with plenty of hiding places. This species does not tolerate constant dampness in the terrarium and likes good ventilation. When these conditions are met, *Anolis roquet caracoli* is not hard to keep and breed.

To keep specimens of *Anolis roquet majolgris*, a keeper should provide an enclosure with robust plants and a relatively high amount of moisture. Also, good ventilation is needed, as well as a temperature of about 26°C/79°F.

PHOTO BY MATTHIAS SCHMIDT.

Anolis roquet majolgris

Description: *Anolis roquet majolgris* is at first sight a very plain anole. The small lizard has a gray-brown ground color that appears lightly mottled on the back through various shades of gray. Only in particularly robust males is the head and nape region extensively marked with white, thereby forming a strong contrast to the rest of the body and the black-ringed eyes. The throat fan of the up to 22-cm (8.8-in) males is about half gray and half yellow. The small (about 11 cm, 4.4 in) females are an inconspicuous gray-brown, and only the often light gray head coloration stands out.

Range and habitat: The anoles are fairly common in the vicinity of the town of Saint Marie in northeastern Martinique. *Anolis roquet majolgris* is a very agile anole and prefers strongly branched, large trees and bushes. Somewhat wetter sites with lush vegetation have particularly high population densities.

Care in the terrarium: *Anolis roquet majolgris* is easy to keep and breed in a medium-sized terrarium. The furnishings should consist of strongly branched plants covering a few robust climbing branches. The lighting should not be too strong, since the anoles prefer partial shade. Good ventilation is indispensable for the well-being of the anoles. *Anolis roquet majolgris* prefers temperatures of about 26°C (79°F) and a humidity of about 70 to 80 percent. The breeding group should consist of either a pair or one male and two females.

Anolis roquet roquet

Description: The ground color of this, the nominate (first-described) form, is green or yellow-green. During courtship the anoles, which can be up to 22 cm (8.8 in) long, reveal the most exquisite colors. Particularly in

Above: Young male example of *Anolis roquet roquet*. **Below:** The females of *roquet* exhibit pronounced dorsal markings.

BOTH PHOTOS BY MATTHIAS SCHMIDT.

the head and nape region, numerous irregularly arranged blue spots and stripes edged by narrow yellow lines stand out from the green ground color. Toward the flanks the mottling often merges into a faint crossbanding. The large throat fan of the male is yellow. Besides the clear difference in size between the sexes (females half the size of the male), the safest distinguishing character is in the throat fans of the anoles. In the males the throat fan extends past the base of the front legs; in the females it ends just before the front legs. Females are light green or brownish green. Characteristic is a beige to dark reddish brown longitudinal stripe on the back. The very small throat fan is pale yellow.

Range and habitat: *Anolis roquet roquet* ranges over all of southern and southwestern Martinique, from the capital of Fort-de-France south along the coast through Trois Ilets to Ste. Luce on the southern coast of the island. The anoles inhabit in large numbers both arid and wetter biotopes. The population density is so high, particularly in parks, that the anoles may be able to control only small territories, and frequently several adult males live in a single palm. Even hedges and fences in the middle of the capital are popular biotopes. The females, which at first sight appear to be in the minority, live rather concealed in the undergrowth. Upon closer inspection, a ratio of 1 male : 2 females or even 1:3 is evident.

Care in the terrarium: *Anolis roquet roquet* is extremely easy to keep. The well-lighted terrarium can be relatively thickly planted and decorated. The anoles need good hiding places as well as suitable basking sites, which absolutely must be dry. A relative humidity between 60 and 80 percent and temperatures of about 27°C (81°F) are necessary for keeping *Anolis roquet roquet* in good health. When the anoles are fed a good and varied diet they are easy to breed. Copulation usually takes place in the morning hours, and about 45 days later the females bury their two eggs in the loose substrate. The young hatch after about six weeks and are easy to rear with tiny insects.

Anolis roquet salinei

Description: These quite large anoles, sometimes more than 23 cm (9.2 in) long, often display their full beauty. On the light green or gray-green ground color, four or five intense black crossbars stand out. Small yellow spots are particularly common in the head and nape region. The head region is yellow in some individuals. The bright yellow throat fan often includes numerous small greenish white markings arranged in lines. *Anolis roquet salinei* is able to change its color, particularly to darker shades. For example, under conditions of high humidity and low temperature the anole turns a very dark gray-brown. The otherwise typical crossbanding then is no longer visible.

The approximately 12-cm (4.8-in) females are light green, gray-green, or gray-brown. The females also have numerous small yellow markings in the head and back region. The throat fan is not present in the females of this subspecies, which makes it somewhat easier to distinguish them from the other *Anolis roquet* females.

Range and habitat: *Anolis roquet salinei* inhabits southeastern Martinique. The anoles occur in the region from Le Marin to Ste. Anne as far as Le Vauclin. The population density is relatively low. The biotope is characterized by extreme aridity, the region being covered with low scrub and with only a few taller trees on which the anoles are somewhat more abundant.

PHOTO BY RALF HESELHAUS.

Males of *Anolis roquet salinei* display rich colors even during the molt.

Care in the terrarium: Of all *Anolis roquet* subspecies, *Anolis roquet salinei* is the form that requires the highest temperatures and intense sunlight. In terrarium keeping, of particular importance are good ventilation and a very arid microclimate. The terrarium should be decorated simply and not be too thickly planted. Small insects, for which *Anolis roquet salinei* shows a strong preference, are particularly suitable for feeding this variety.

Anolis roquet summus

Description: *Anolis roquet summus* is without doubt one of the most beautiful anoles of the *roquet* group. At a body length of about 22 cm (8.8 in), the dark emerald-colored males are fully grown. Particularly in the head and nape region, which can also exhibit a yellow ground color, the anoles have numerous white spots with blackish blue borders. In rare cases the dark blue color can be extensive and extend as far as the hind legs. The contrasting markings of the blue-black, the intense green ground color, and the orange-yellow throat fan are particularly striking in *Anolis roquet summus.* Adult females are easy to distinguish from the females of other *Anolis roquet* subspecies. The very robust anoles have a green ground color, and the head region, down to the jaws, is marked with black or dark gray. This black color continues in the form of two broad dorsal stripes extending onto the tail. On the neck and flanks the anoles can exhibit a slight blue suffusion. The lower jaw and throat are yellowish mixed with a light blue tint.

Females of *Anolis roquet salinei* are characterized by a fine speckling.

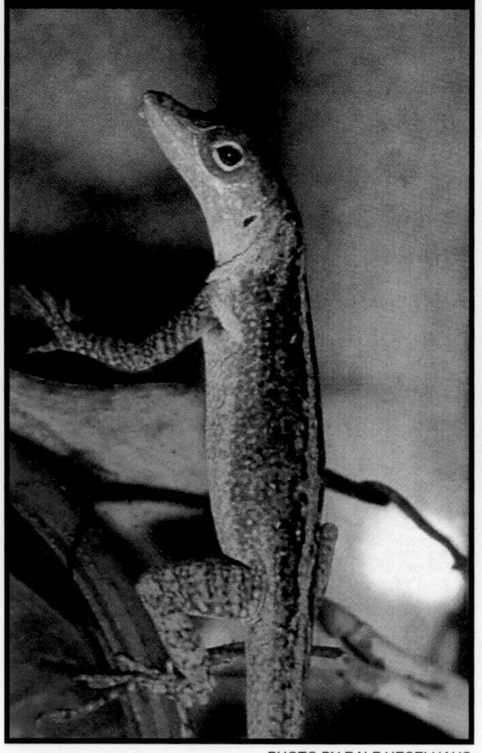

PHOTO BY RALF HESELHAUS.

Range and habitat: *Anolis roquet summus* is exclusively an upland form and inhabits the northern part of Martinique at altitudes between 400 and 900 meters. The lush, dense vegetation found there is extremely rich in species, large trees covered with epiphytes underscoring the overall impression of a tropical rainforest. The humidity is very high at 80 to 100 percent. Temperatures are about 25°C (77°F) during the day and drop to below 20°C (68°F) at night. The population density is highly variable and reaches its maximum near small villages.

Care in the terrarium: In contrast to all other anoles on Martinique, *Anolis roquet summus* tolerates high humidity, even over long periods of time. This means that this subspecies is an excellent choice for a purely tropical terrarium. To be sure, at times dry areas should be available, but this is not a prerequisite for good breeding success. *Anolis roquet summus* is a particularly well-suited terrarium animal and will present no problems in keeping and breeding.

PHOTO BY MATTHIAS SCHMIDT.

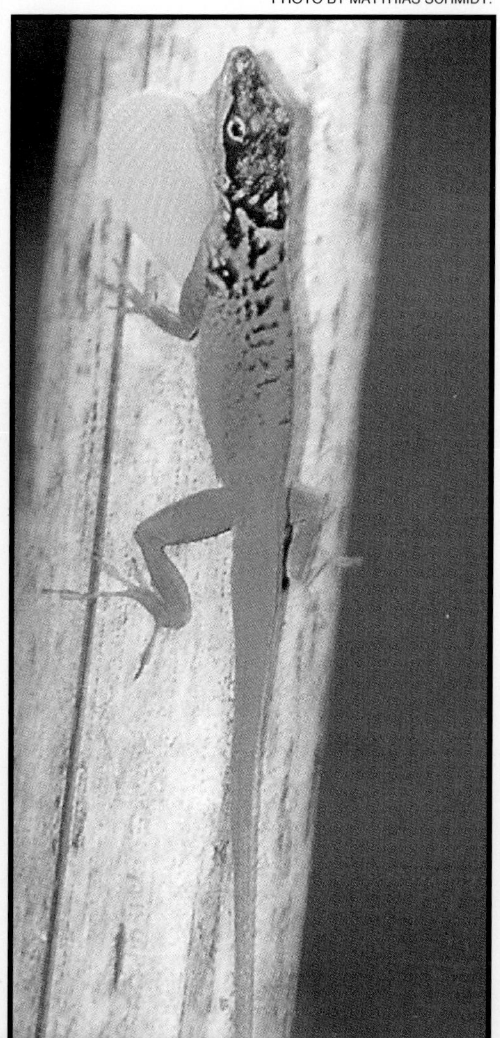

Unlike the other anoles on the island of Martinique, *Anolis roquet summus* can tolerate a high level of humidity. It is known to make a hardy and rewarding captive.

Anolis roquet zebrilus

Description: *Anolis roquet zebrilus*, which grows to a length of more than 23 cm (9.2 in), is characterized by a gray-brown ground color throughout that can be somewhat lighter in the head region and exhibits small black markings. The light gray coloration of the throat extends about halfway over the large throat fan. The other half is light yellow with small white patches. The name of this subspecies refers to the characteristic black crossbanding on the nape and back. As a result of the dark crossbanding, these anoles closely resemble *Anolis roquet salinei*, only the gray ground color permitting the quick identification of the two subspecies. The up to 14-cm (5.6-in) females often have darkly mottled, faint crossbands on the back on a gray-brown ground color. The small throat fan is dark orange.

Range and habitat: *Anolis roquet zebrilus* inhabits the northwestern coast of Martinique in the region of the towns of St.-Pierre and Le Carbet. The anoles are found primarily in large trees, among ruins, and on the walls of houses, which also provide a favorite place to bask. Because of the location of the range on the leeward side of the mountains, the biotope is relatively arid.

Care in the terrarium: In keeping with the arid and sun-drenched habitat, *Anolis roquet zebrilus* needs a well-lighted and well-ventilated terrarium. Accordingly, the cage should not be thickly planted. The daytime temperatures can reach maximum values of about 30°C (86°F), and the nocturnal temperature should be 24°C (75°F). The anoles do not react very sensitively to high humidity, but based on experience the relative humidity should not exceed 75 percent.

Anolis sabanus, Panther Anole

Description: *Anolis sabanus* formerly was considered to be a subspecies of *Anolis bimaculatus*, but based on more recent interpretations it is viewed as an independent species. The up to 18-cm (7.2-in) anole exhibits a conspicuous leopard pattern of large blackish brown blotches that contrast with the silvery gray to light brown ground color. The leopard markings continue more or less distinctly onto the legs. The male's throat fan is pale yellow.

Range and habitat: The Panther Anole occurs only on Saba, an island in the Lesser Antilles formed by the crater of an extinct volcano. Steep cliffs up to 600 meters high characterize the landscape of this

PHOTO BY MATTHIAS SCHMIDT.

Anolis roquet zebrilus sometimes is found around human residences, where it often basks on the walls of houses. In the terrarium, specimens need very little plant cover, a daytime temperature of about 30°C/86°F, and a 75% humidity.

ARTWORK BY JOHN R. QUINN.

Anolis sabanus, Panther Anole.

island. Lush green valleys are found only in the interior of the island. *Anolis sabanus* lives there on thick tree trunks, while near the coast it lives on a rocky substrate.

Care in the terrarium: The relatively small Panther Anoles can be kept in pairs in cages with dimensions of at least 70 X 50 X 50 cm (28 X 20 X 20 in). The back wall of the terrarium should be formed as a rock wall, because this species likes to climb on rocks. To provide shelter and an additional place to climb, a bizarrely shaped piece of driftwood or something similar can be placed in the foreground and then planted with bromeliads (for example, *Vriesea*). The air temperature should vary between 22 and 30°C (72 and 86°F), with moderate humidity.

The females lay their clutch of two eggs several times a year in the soil or in the mulm of old rootstocks. The maturation of the clutch is best carried out in an incubator. At a temperature of 26°C (79°F), the young hatch after 40 to 50 days. Because of their pronounced aggressiveness, rear them individually in small plastic terraria. During rearing, a varied diet with regular supplements of a calcium and vitamin preparation is necessary.

Anolis sabanus is endemic to Saba, an island in the Lesser Antilles formed by the crater of an inactive volcano.

PHOTO BY RALF HESELHAUS.

Anolis sagrei, Brown Anole

Description: *Anolis sagrei* has been one of the most popular anoles in the terrarium for years. Males have a chocolate ground color with dark brown diamonds along the backbone and a pattern of beige-yellow dots. The females also are brown, but have a light yellow dorsal stripe and two brown stripes on the sides. At a length of about 14 cm (5.6 in), they are considerably smaller than the up to 18-cm (7.2-in) males. *Anolis sagrei* is one of the "high-headed" species: it does not have the flat alligator snout typical of many anoles, such as *Anolis carolinensis*. The males are characterized by a large reddish brown throat fan with black streaks and a yellow margin. The small, non-serrated nape crest is revealed only when the anole is excited.

Range and habitat: Originally *Anolis sagrei* was found mainly on Cuba. From there it colonized Florida

BOTH PHOTOS BY GUIDO DINGERKUS.

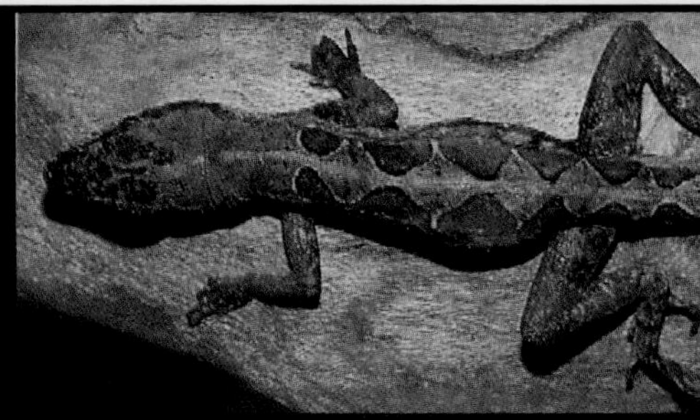

Brown Anole, *Anolis sagrei* ssp., showing the lateral and dorsal patterns.

and possibly also has been introduced to Jamaica. Additional subspecies live in the Bahamas *(Anolis sagrei ordinatus)*, on Cayman Brac *(Anolis sagrei luteosignifer)*, and on Swan Island *(Anolis sagrei nelsoni)* off the Yucatan. The nominate subspecies *(A. sagrei sagrei)* also occurs on the Mexican coast in the Yucatan and in Belize. In addition to *A. sagrei sagrei* over most of Cuba, a second subspecies, *A. sagrei greyi*, occurs in central Cuba. In Cuba we found *Anolis sagrei* on low bushes along roads and paths. In particular, females often went to the ground to reach the nearest bush.

Care in the terrarium: *Anolis sagrei* can be kept in pairs in medium-sized terraria. The males are extremely aggressive toward one another, but in large terraria it is

ARTWORK BY JOHN R. QUINN.

Above: *Anolis sagrei ordinatus.*
Below: *Anolis sagrei greyi.*

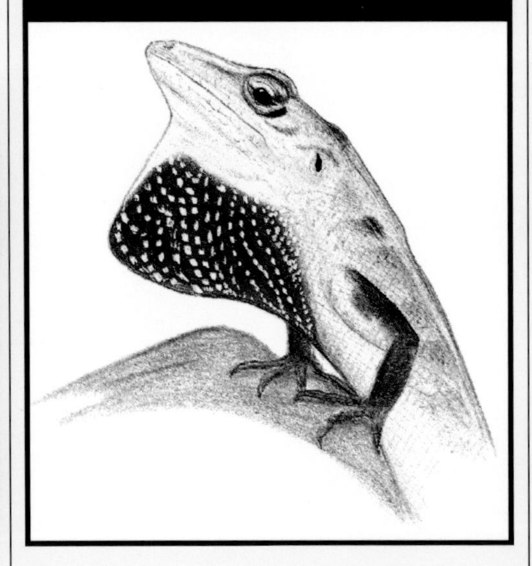

A beautiful example of a young male Brown Anole, *Anolis sagrei.* Some refer to this as the animal's "russet" phase.

PHOTO BY R. D. BARTLETT.

PHOTO BY MATTHIAS SCHMIDT.

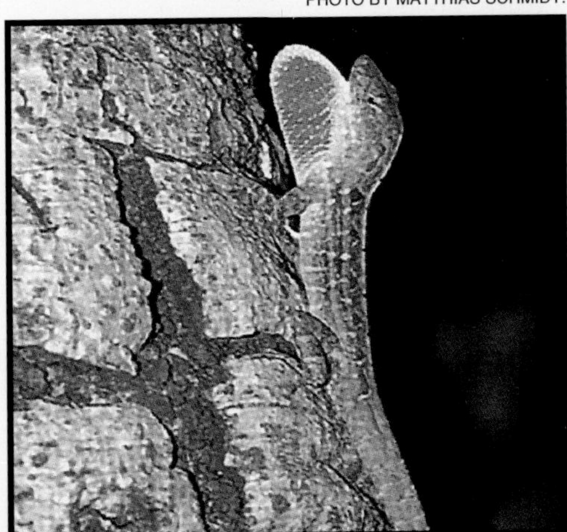

A courting male Brown Anole, *Anolis sagrei*, from an introduced population in Florida.

possible to keep the Brown Anole with "green species" like *Anolis marmoratus* and *Anolis speciosus. Anolis sagrei* requires moderate humidity and temperatures of 25 to 28°C (77 to 82°F). Hibernation from November to February at lowered temperatures and reduced rations is recommended. During the breeding season the female buries a clutch of two eggs in the soil about every four weeks. Maturation takes place in the incubator. At a temperature of 25°C (77°F), the young hatch after 50 to 60 days.

Anolis sagrei scriptus.

ARTWORK BY JOHN R. QUINN.

Anolis trinitatis

Description: *Anolis trinitatis* is an extremely attractively marked anole. The ground color of the up to 18-cm (7.2-in) males is an emerald green. The head region is a striking sky-blue, and the eyes have dark blue rings. Depending on mood, the anoles can also exhibit yellow heads. The proximal third of the tail is yellow-green and ends in dark blue or black. The legs are dark green. The throat fan is rich yellow and can exhibit white patches. The smaller (13 cm, 5.2 in) and more slender females are uniformly green above,

ARTWORK BY JOHN R. QUINN.

Anolis trinitatis.

the belly as well as the small throat fan yellowish white.

On St. Vincent there are various color varieties of *Anolis trinitatus* that, in part, clearly differ from the above-described population from the Kingstown Botanical Garden. One of the most attractive color forms is found in the Mesopotamian Valley. The ground color is dark blue, broken up only by small dark green and yellow patches on the head, nape, and legs. The coloration of the light blue throat fan, which features numerous small yellow and red dots, is repeated on the legs and tail.

Range and habitat: Fifteen color varieties of *Anolis trinitatis* occur on the Caribbean island of St. Vincent. The center of distribution is the region around the capital of Kingstown, in particular in the Botanical Garden, and the fertile Mesopotamian Valley in the southeastern part of the island. The anoles prefer to stay in the lower regions of medium-sized trees and in low but not densely branched shrubs. The temperature in the predominantly shady biotopes is about 27°C (81°F). Only during the dry season from January to May do temperatures rise above 32°C (90°F).

Care in the terrarium: Even relatively small terraria with dimensions starting at 50 X 50 cm (20 X 20 in) are suitable for *Anolis trinitatis*. The anoles, which if possible should be kept in pairs, are easy to keep and breed. The terrarium can be thickly planted and decorated with roots, but should have a few open places to bask. The humidity should be kept between 60 and 90 percent. Pregnant females lay their two eggs about 25 days after copulation, finding a suitable site in the loose substrate. As a rule, the females repeatedly use the same laying site, making the clutch easy to find. The about 5-cm (2-in) young hatch after 35 days and can be reared easily with *Drosophila*. Calcium and vitamin supplements as well as regular ultraviolet irradiation are essential for the further growth of the young *Anolis trinitatis*.

A remarkably attractive male example of *Anolis trinitatis*.

PHOTO BY RALF HESELHAUS.

OTHER CARIBBEAN ANOLES

The following list represents the Caribbean anole species generally recognized as valid. Several other taxa (listed after the species that have been more fully treated) have been elevated to species rank from subspecies, are poorly known, or recently were described by various workers. Though it is not unlikely that they do deserve species status, they are not detailed here. See Schwartz and Henderson, 1991, *Amphibians and Reptiles of the West Indies. Descriptions, Distributions, and Natural History* (Univ. Florida Press), for a full discussion of all the species of the Antilles and lists of their subspecies. Additional species and subspecies probably will have been described by the time you read this book, so no list of Caribbean anoles can be complete for very long.

In the listing, each species is briefly mentioned along with its general distribution, length of males, and a summary of the colors and patterns. Be aware that many of the species have two or more subspecies that may differ considerably in coloration. No one volume yet exists to allow the identification of all Caribbean anoles, but keys to many of the species are presented in Henderson and Schwartz, 1984, "A guide to the identification of amphibians and reptiles of Hispaniola," *Spec. Publ. Milwaukee Public Mus. Biol. and Geol.*, 4: 1-70; and Schwartz and Henderson, 1985, *A guide to the identification of amphibians and reptiles of the West Indies exclusive of Hispaniola*, Milwaukee Public Mus., 165 pages.

THE MAJOR SPECIES

Anolis acutus. Virgin Islands, St. Croix. 20 cm (8 in). Back bronze, varying through gray to dark brown; throat fan white to dull gray with yellow or orange margin.

Anolis aeneus. Grenada, Grenadines. 21 cm (8.4 in). Gray ground color, blue suffusion in head and nape region, dark green vertical bars on the back, blue eye ring; female longitudinally striped.

Anolis ahli. Cuba. 17 cm (6.8 in). Ground color brown with suggestions of red striping, black patch in nape region; throat fan yellow with large red spot.

Anolis allisoni. Cuba. 22 cm (8.8 in). Ground color green, head and shoulders bright blue; throat fan ruby.

Anolis allogus. Cuba. 19 cm (7.6 in). Dorsal coloration reddish brown with yellow net-like markings; distinct tail crest; throat fan yellow with three vertical stripes.

Anolis anfiloquioi. Cuba. 12 cm (4.8 in). Black spot in neck region; throat fan orangish gray.

Anolis angusticeps. Cuba, Isla de la Juventud, Bimini, Antigua, Berry Islands, New Providence, Eleuthera, Exuma Cays, Long Island, Cat Island. 16 cm (6.4 in). Dorsal coloration grayish brown or yellowish brown, occasionally dark brown; throat fan orange to yellow.

Anolis argenteolus. Cuba. 16 cm (6.4 in). Ground color brown with yellow or gray net-like markings; throat fan white with a gray or brown area.

Anolis argillaceus. Cuba, Isla de la Juventud. 14 cm (5.6 in). Dorsal coloration with grayish brown net-like markings; large yellow-orange or red throat fan.

Anolis baracoae. Cuba. 50 cm (20 in). Shoulder stripe consisting of yellow spots ringed with black; throat fan pink, blue-green, white, or brownish with green spots.

Anolis bremeri. Cuba, Isla de la Juventud. 21 cm (8.4 in). Throat fan ocher or mahogany.

Anolis brunneus. Crooked-Acklins Islands, West Plana Cay, Bahamas. 23 cm (9.2 in). Ground color grayish brown, black oval blotches behind the eyes and above the front leg; ruby throat fan.

Anolis centralis. Cuba. 14 cm (5.6 in). Ground color white or light brown; throat fan yellow to red.

Anolis clivicola. Cuba. 15 cm (6 in). Light green ground color with small dark green dots and gray dorsal stripe, the sides dark red; throat fan violet.

Anolis conspersus. Grand Cayman. 21 cm (8.4 in). Ground color mottled green or brown; throat fan blue.

Anolis cristatellus. Puerto Rico, Virgin Islands, Dominican Republic. 23 cm (9.2 in). Dorsal coloration grayish green; throat fan varies between red and green.

Anolis cupeyalensis. Cuba. 13 cm (5.2 in). Ground color intense green with a pink dorsal stripe; throat fan green.

Anolis cybotes. Hispaniola, introduced to Florida. 18 cm (7.2 in). Background color tan to gray, dark dumbbells down back; throat fan pink to yellow or dirty white.

Anolis delafuentei. Cuba. 18 cm (7.2 in). Ground color brown; throat fan red or orange.

Anolis desechensis. Isla Desecheo off Puerto Rico. 17 cm (6.8 in). Ground color grayish brown; throat fan yellowish orange.

Anolis distichus. Bimini, Andros, Berry Islands, New Providence, Eleuthera, Exuma Cays, Cat Island, Long Island, San Salvador, Rumy Cay, Ragged Islands, Haiti, Dominican Republic, Florida. 16 cm (6.4 in). Ground color yellowish brown to gray; throat fan yellow or orange.

Anolis ernestwilliamsi. Virgin Islands. 25 cm (10 in). Dorsal coloration grayish brown with a black spot; throat fan rich red with green center.

Anolis evermanni. Puerto Rico. 21 cm (8.4 in). Dorsal coloration intense green or brown; throat fan yellow.

Anolis fairchildi. Cay Sal Bank. 22 cm (8.8 in). Dorsal coloration green, sides yellowish; pointed head; throat fan bicolored, top half white, bottom half ruby with white spots.

Anolis fugitivus. Cuba. 11 cm (4.4 in). Dorsal coloration gray with beige dorsal stripe, head green; throat fan pale yellow.

Anolis grahami. Jamaica. 17 cm (6.8 in). Head and body green, base of tail blue, tail turquoise; throat fan yellowish orange.

Anolis guazuma. Cuba. 14 cm (5.6 in). Throat fan yellow, edged in orange; body length longer than tail.

Anolis gundlachi. Puerto Rico. 18 cm (7.2 in). Ground color dark olive with three to five black crossbands; tail crest; throat fan orange.

Anolis homolechis. Cuba, Isla de la Juventud. 21 cm (8.4 in). Ground color brown with white lateral stripe; throat fan white to gray.

Anolis imias. Cuba. 19 cm (7.6 in). Dorsal markings with five dark crossbands, chin with black net-like markings.

Anolis isolepis. Cuba. 12 cm (4.8 in). Ground color green, back with purple net-like markings; throat fan yellow or white.

Anolis juangundlachi. Cuba. 10 cm (4 in). Grayish brown ground color; throat fan pale yellow.

Anolis jubar. Cuba. 18 cm (7.2 in). Ground color brown to gray; throat fan yellow to orange; subspecies with three white dorsal

APM
Introduction to Programme Management

Paul Rayner

on behalf of
ProgM, the Programme Management Specific
Interest Group of the APM

Association for Project Management

Association for Project Management
150, West Wycombe Road
High Wycombe
Buckinghamshire
HP12 3AE

British Library Cataloguing in Publication Data is available
ISBN 10: 1-903494-63-X
ISBN 13: 978-1-903494-63-9

Cover design by Fountainhead
Typeset by RefineCatch Limited
Printed by Latimer Trend and Company
Copy editor: Merle Read
Proofreader: Patrick Cole
Publishing Manager: Ingmar Folkmans

CONTENTS

FOREWORD BY MIKE NICHOLS, CHAIRMAN OF THE ASSOCIATION FOR PROJECT MANAGEMENT

I am convinced that programme management has the potential to offer significant improvements to many commercial and public sector organisations. Yet whilst many organisations are able to manage projects effectively, fewer can be sure they are investing in the optimum portfolio of work. Organisations may believe that their projects are being 'done right', but the confidence that they are 'doing the right projects' may be lacking. Where correctly implemented, programme management will provide that confidence.

Like many relatively new topics, programme management is surrounded by some confusion and curiosity. To provide clarification and routes to further knowledge, the APM has produced this introductory booklet.

Whilst recognising that the terms *programme management* and *programme* have differing meanings and connotations across the commercial and public sectors, both in the UK and elsewhere, this introduction provides a view that is supported by the APM and aims to establish a standard for members and supporters of the organisation.

Mike Nichols

ACKNOWLEDGEMENTS

Project Sponsor: Mike Nichols, Chairman of the APM

Project Manager: Geoff Reiss, Geoff Reiss Limited

Lead Author: Paul Rayner, Chairman, ProgM

The author and the APM wish to thank those people that took the time and trouble to read and comment on early drafts of this publication, including, in alphabetic order: Adrian Pyne, Andrew Scheiner, Clive Savory, David Partington, Elizabeth Hirst, Geraldine Duffy, Ian Koenig, John Chapman, Leif Hertz, Neil Cockshaw, Penny Pullen, Pete Corcoran, Peter Simon, Ruth Murray-Webster, Stefan Hallberg and Tracey Marshall.

Guidance on the role and objectives of ProgM can be found in the Appendix.

1
Programmes and programme management

1.1 WHAT IS A PROGRAMME?

Over the last 15 years, the growing scale of change, the need to respond quickly to changing business environments and the impact of new technologies have led many organisations to adopt programmes as the means of achieving organisational change. Programmes are temporary management structures designed to help organisations to achieve specific objectives. As such, they come in all shapes and sizes, and the term 'programme' is applied to many different structures. Thus the use and meaning of the term can vary widely across industry sectors and business cultures.[1]

The *APM Body of Knowledge*[2] defines programme management as: 'the co-ordinated management of related projects, which may include related business-as-usual activities, that together achieve a beneficial change of a strategic nature for an organisation'. Other definitions of programme management are available from bodies such as the Office of Government Commerce (www.ogc.gov.uk) and many public sector and commercial organisations (see also section 5).

Whilst there are many variations in the way that the concepts of programme management are applied, the programmes that result display the following characteristics:

- their purpose is to deliver the capability to make strategic, significant or step changes to organisations, normally referred to as or measured by benefits;[3]

---

[1] In some countries the US spelling – program – is normally used, whereas in Britain this spelling usually refers to the code that drives a computer (e.g. a 'computer program').
[2] *APM Body of Knowledge*, 5th edn (2006), available from the Association for Project Management (see section 5 for details).
[3] Sometimes benefits are referred to as outcomes or as business benefits. For brevity, the term 'benefit' will be used throughout this document. Further guidance on benefits and their management can be found in section 3.2 of this document.

- this need for significant improvement will be consistent with the organisation's strategy, and programmes will help to deliver elements of that strategy;
- the creation of the desired benefits will be achieved only through the co-ordination and successful completion of a number of component projects;
- different parts of an organisation or differing organisations may be affected by the programme;
- the overall measure of success will be determined by the actual delivery of the expected benefits, which frequently involves the use of capabilities or facilities created by the programme in an ongoing, 'business as usual' manner.

As an introduction to the subject, this document will adopt a simplified definition, i.e. that a programme is: 'a co-ordinated set of projects that together achieve a beneficial change of a strategic nature for an organisation'.

1.2 WHAT IS PROGRAMME MANAGEMENT?

Because programmes are the method by which strategy is delivered, programme management provides a management interface between those responsible for deciding strategy and those responsible for managing the component projects and other activities. Thus, typical programme management responsibilities include:

- interpreting the organisation's strategy so as to create practical programmes of improvement and change that will be successful within the organisation's cultural and economic environment;
- selecting, initiating and monitoring the component projects that make up the programmes, including defining the scope of individual projects;
- co-ordination between component projects, so as to maximise the value of the combined deliverables of the component projects into fully usable capabilities that may be used to deliver the desired benefits;

- cancelling projects or changing the scope of projects in reaction to changes to the organisation's strategy or environment;
- identifying, supporting, measuring and monitoring the actual delivery of benefits.

In summary, programme management provides a layer of management, above that of the component project management teams, focused on defining, integrating and co-ordinating the projects in order to deliver benefits and to realise strategy (Figure 1.1).

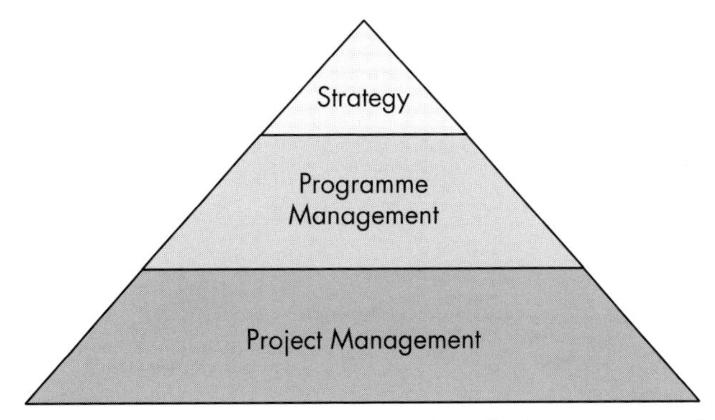

Figure 1.1 *Programme management in relation to organisational strategy and to project management*

1.3 PROGRAMME MANAGEMENT APPLICATION

Project management is designed to focus attention on the delivery of specific outputs: the creation of deliverables or products against agreed time, quality and cost targets. By contrast, programme management is designed to focus attention on outcomes: improvements or beneficial change to an organisation.

Therefore the organisations that will benefit most significantly from the application of programme management techniques will normally be those seeking to deliver beneficial change to the organisation in line with a defined strategy.

1.4 TYPES OF PROGRAMME

As the mechanisms by which organisations deliver their strategies, programmes are as varied as the organisations that initiate them and the strategies they seek to fulfil. Nevertheless, research by ProgM suggests a strong focus on organisational change, often underpinned by IT. For example, as shown in Figure 1.2, programmes analysed by the Programme Management Maturity Model[4] show that over 50 per cent are classified by those working on them as being programmes with a significant IT element. A further 20 per cent classify their programmes as being focused on organisational change.

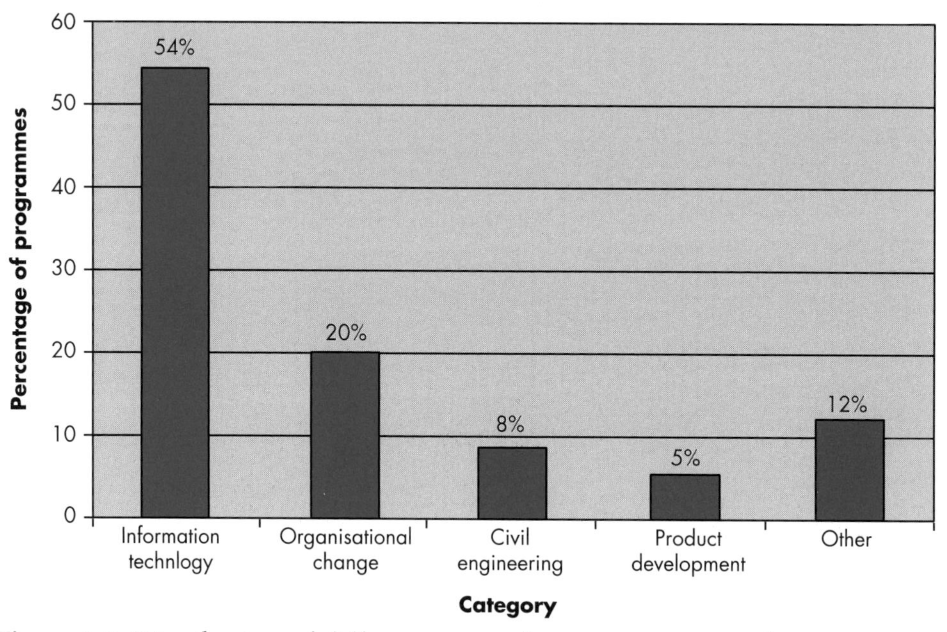

Figure 1.2 *Distribution of different types of programme according to purpose*

The heavy focus of programmes on IT and organisational change is reflected in the popularity of programme management in industry sectors such as finance and government, as shown in Figure 1.3.

[4] The Programme Management Maturity Model (PMMM) is a tool to analyse the maturity of programmet processes. Its use is free and it allows those responsible to compare their programmes against a database of approximately 100 programmes of all types from all parts of the world. Details of the PMMM are available from the ProgM website (www.e-programme.com/pmmm.htm).

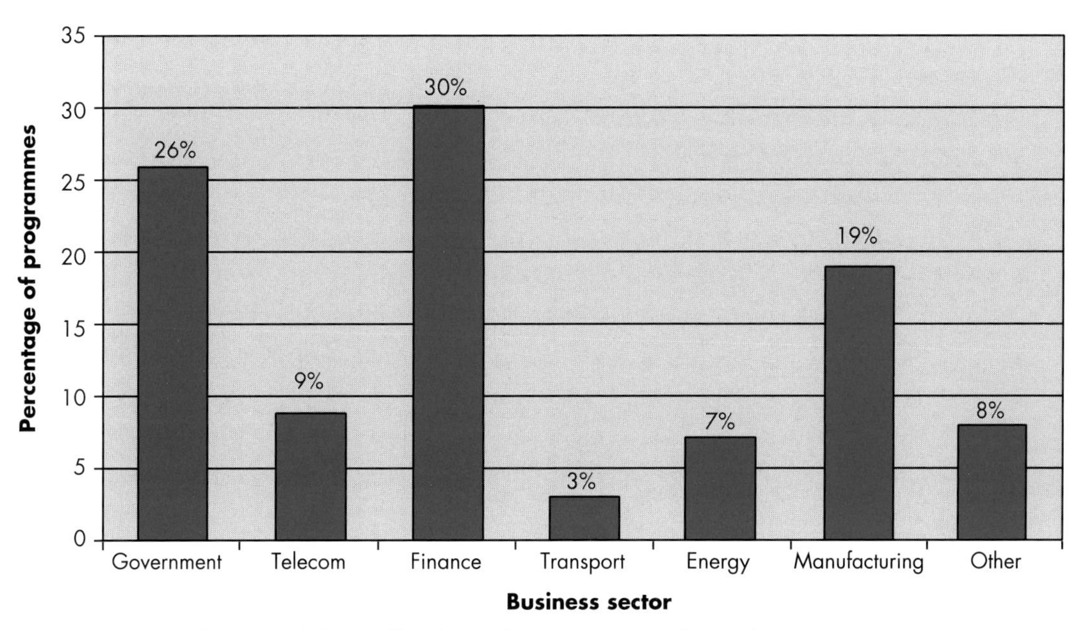

Figure 1.3 *Distribution of programmes by industry sector*

These programme types can also usually be thought of as business change or transformation programmes, in that they seek to change some aspect of an organisation, or even the organisation itself.

1.5 HOW PROGRAMMES DIFFER FROM PROJECTS

Whilst there is frequently overlap between programme and project management activities, it is wrong to regard programmes as merely large and complex projects. They are usually larger than projects but not necessarily so.

ProgM research in the UK[5] suggests that projects, on average, involve 17 people and can be expected to last over 20 months, whereas programmes typically involve 167 people and are expected to last over 40 months. However, the key difference is that programmes have a different

[5] See 'Analysis of Research Questionnaire', October 2005, available on the ProgM website at www.e-programme.com.

purpose and require different management structures and skills to be successful.

Projects are the means to deliver specific one-off deliverables. To be successful, the required deliverables must be defined and specified in advance, with defined budgets and timetable expectations. By contrast, programmes are the means to deliver benefits or outcomes, and amongst their activities are those needed to define and agree the scope of the various projects that will make the achievement of the desired benefits possible.

For example, a project might create a new warehouse, i.e. a deliverable. A warehouse on its own may seem to have little direct value, but when it is combined with the deliverables of other projects – such as a computerised stock-control system, a retrained workforce, a new organisational structure, or a new staff bonus scheme – in a programme, it can provide the capability of supplying customers faster, with reduced costs and less wastage due to goods damaged in transit.

Success for a project is usually defined as creating the required deliverables to an adequate standard, within agreed time and cost constraints. Whether the deliverable, such as a new warehouse, is successfully used or not is not the point. Indeed, there are many projects that have been deemed highly successful that have created deliverables that have never actually been used. Success for a programme is usually measured in terms of creating a whole new capability and, increasingly, the extent to which the expected benefits are actually realised.

The term programme management is often incorrectly used to refer to the execution of a number of projects by a contractor for a client. Examples include the construction of a number of water treatment plants for a local utility body and a series of retail unit refurbishments by a shopfitting contractor for a retail chain. Such multi-project management activities often share a common budget and resources; frequently there will be a common approach and methodology.

However, this work will typically form a part of a larger programme, and it is only at that programme level that benefits will play a key role. In the case of the water treatment plants, other projects to measure water quality and improve other parts of the infrastructure will combine with the treatment plants to deliver the key performance indicators required by the industry regulator, OFWAT. In the retail example, only when the marketing, management information systems and staff training projects complement the shop refurbishments will the increased income and customer satisfaction benefits become possible.

Table 1.1 summarises some of the key differences between programmes and projects.

Table 1.1 *Differences between programmes and projects*

Aspect	Programmes	Projects
Clarity of scope	Programmes involve uncertainty in funding, range and impact	Projects require clearly defined scope, budget and timescales
Clarity of deliverables	Specific deliverables to be created are usually unclear at the start	The required deliverables are usually clearly defined at the start
Structure	Separately managed projects, which must be co-ordinated. This structure may be unclear at the start and may change throughout the life of the programme	Projects form a single managed entity, which is clear at the start and will not usually change significantly during the life of the project
Methodologies or approaches	Frequently involves co-ordinating and managing several different organisations, each of which is responsible for one or more discrete projects, and each of which may be using a different methodology or project approach	A single project is normally the responsibility of a single organisation, working to a single methodology or project approach
Clarity of budgets and timescales	At the start, the time and budget required will often be unclear, and part of the role of the programme will be to define these	Projects start with a project initiation document, project management plan, business case or equivalent that defines expected costs and timescales
Approach to change	Because the scope and deliverables are unclear, change to priorities and requirements is constant and a major feature of programmes	Changes to scope or desired deliverables are generally unwelcome and subject to rigorous control
Critical activities	A major element is managing people and organisational issues necessary to ensure that the new capabilities will be used to deliver the desired benefits	The major element is managing the technology or specialist skills necessary to create the deliverables
Measure of success	The creation of usable capability and/or the delivery of business benefits	The creation of the specified deliverables within agreed time and cost constraints

1.6 HOW PROGRAMME MANAGEMENT DIFFERS FROM PORTFOLIO MANAGEMENT

Since the early part of this decade, there has been growing interest in portfolio management. This has many similarities to programme management, not least its focus on the achievement of benefits. However, whereas programme management relates to the co-ordinated management of a set of related projects – typically where the projects are mutually dependent and all are required to create the required capability and business benefits – portfolio management is generally applied to unrelated projects.

The linking characteristics of a portfolio lie in areas other than benefits. The common factor is usually that all of the projects lie within the same organisation or department, and they must be financed through a common source of funding or they need to make efficient use of a common pool of resources. Thus portfolio management is akin to the similarly named activity which takes place in the financial world, where a portfolio of investments is managed to yield maximum returns, capital growth and so on. In such a portfolio there is no relationship between the different stocks and shares – each investment is self-contained and is bought or sold only to achieve the objectives of the portfolio as a whole.

Portfolio management is an approach that is widely used in the IT and civil engineering sectors, and often enables more effective management and support for IT projects – for example, it helps ensure the efficient use of development resources, such as business analysts, solutions architects, web designers, and so on – whilst minimising the costs through the elimination of duplicate management and support activities. It also helps to create an understanding of how the various IT projects will contribute or not to the achievement of strategies of the various business units for which they are being run – something that is not always clear with traditional approaches to project management.

Portfolio management may be characterised as a 'mass production' approach to project management, ensuring that the portfolio as a whole meets the organisation's objectives, with projects being added or removed independently of others in the portfolio. Programme management is closer to 'batch production', where the component projects are mutually dependent and should only be added to or removed if that will contribute to the better achievement of that programme's benefits. It should also be noted that a portfolio may include programmes as well as projects.

The key fact is that both programme and portfolio management require a similar strategic awareness to be successful, similarly skilled people and similar organisational empathy, similar flexibility and a similar focus on benefit achievement.[6]

1.7 HOW PROGRAMMES DELIVER BENEFITS

The role of the programme in actually delivering the benefits may vary. Wherever possible, the programme should seek to realise measurable benefits as soon as possible, and benefits are frequently realised during the life of a programme. However, most benefits will be realised only through the ongoing, 'business as usual' use of the new facilities or capabilities created through the programme.

Where benefits are delivered after the programme team has been disbanded or assigned to new endeavours, responsibility for monitoring, measuring and realising the benefits must be transferred to an appropriate function.

The relationship between projects, programmes and benefit delivery is thus in accordance with Figure 1.4, which shows how the deliverables created by projects are combined by programmes to create capabilities, which are then used so as to realise benefits. Benefits management ensures the realisation of benefits, and responsibility for it may rest partly with the programme management team and partly with another group, such as the main board or the organisation's finance function.

A critical component within benefit management is the project(s) or set of activities needed to manage the transition from the old ways of working, based on previous processes, tools and capabilities, to the new ways.

[6] For a definition and further guidance on portfolio management, see the *APM Body of Knowledge*.

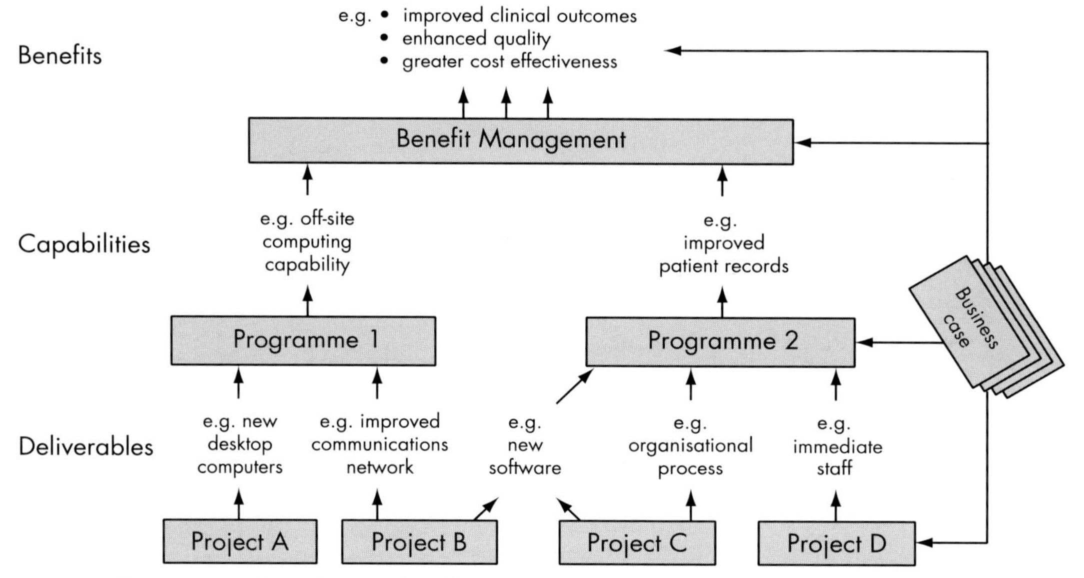

Figure 1.4 *The relationship between projects, programmes and benefits*

Such transitions are frequently very complex and risky. For example, they may involve the switch-off of old facilities and the switch-on of new ones according to tight timescales. Often they depend upon the commitment of users and line management who are not directly under the control of the programme or any of its component projects. Furthermore, there can be little allowance for failure or delay since service to customers must continue throughout the transition and, unless a smooth transition is effected, it may not be possible for the organisation to take advantage of the new capabilities and thus be able to realise the benefits that the programme is intended to provide.

1.8 PROGRAMME MANAGEMENT AND STRATEGIC DIRECTION

It is recognised that senior managers and executives must deal with uncertainty and ambiguity. Strategic planning and setting the direction for an organisation is fundamentally different from operational management. By contrast, project managers desire certainty, since in the definition of a project there should be clarity of purpose, cost, timescale and quality targets. When project managers work with directors these differences can cause friction as the expectations of both parties can be fundamentally different.

Programme management teams fill this gap between directors and project managers. Their skills are in understanding the strategic direction of the organisation, ensuring that there is an alignment of the portfolio of projects to support the business, working in an uncertain environment and responding to change.

The organisation's directors can then concentrate on setting direction, considering short- and long-term issues. The programme manager will then ensure that this is translated into the language of projects, manage the project managers and deliver to the organisation the required changes and capabilities.

In many organisations change programmes tend to cut across business-as-usual structures. For example, a programme transforming a bank's operation to Internet-based services will need to interact with the bank's existing vertical and functional structures, i.e. operations, IT, human resources (HR), finance, marketing and so on. Yet the aims and objectives of these groups may not always be aligned and, unless such interactions are carefully planned in conjunction and with the co-operation of each business unit, the programme could run into a 'brick wall' of non-cooperation. Planning and managing such interactions are a key activity within programme management.

Most large organisations have several change programmes running concurrently. Therefore the most senior levels of management need to take seriously not only the 'sponsorship' roles for individual programmes, but also the management of the portfolio of change programmes. An organisation's directors must create the environment in which change programmes can succeed in delivering outcomes and benefits, and hence the strategy.

Business change and programme management should therefore be well understood by the most senior management in an organisation and be represented at that level.

1.9 OTHER INTERPRETATIONS AND USES OF THE TERMS

In the construction industry the term 'programme' often refers to the timetable of activities that must be completed, often presented in the form of a Gantt chart, whilst 'programme management' refers to the process of integrating separate project schedules. For example, on a large engineering project there may be several contractors, each managing a range of subcontractors – all of whom will produce their own separate schedules of work, referred to as 'programmes' – and the integration of these many schedules into a coherent master schedule would be called 'programme management'.

Also in the construction and heavy engineering industries the term 'programme management' is often used by contracting organisations to refer to a portfolio of projects that benefit from a consistent or integrated form of management. These projects typically result in deliverables created by a contractor for a client organisation in exchange for payment, and therefore the contractor has a limited interest and influence over the delivery of benefits.

2
The programme life cycle

2.1 A HIGH-LEVEL PROGRAMME MANAGEMENT LIFE CYCLE

Programmes can vary enormously in size. Analysis of programmes submitted for review by the Programme Management Maturity Model (see section 1.4) show programmes ranging from small, made up of a handful of projects and employing in total a team of 20 people, through to giants made up of 50 or more projects employing more than 2000 people. In spite of the enormous variation in size of programme, all can be deemed to follow a standard life cycle, the key elements of which are shown in Figure 2.1.

This is a very high-level view of the programme life cycle. Because of the large scale of many programmes, they are often subdivided into tranches or phases, especially in the execution phase. This does allow more time to define projects in later tranches and for early benefits to be

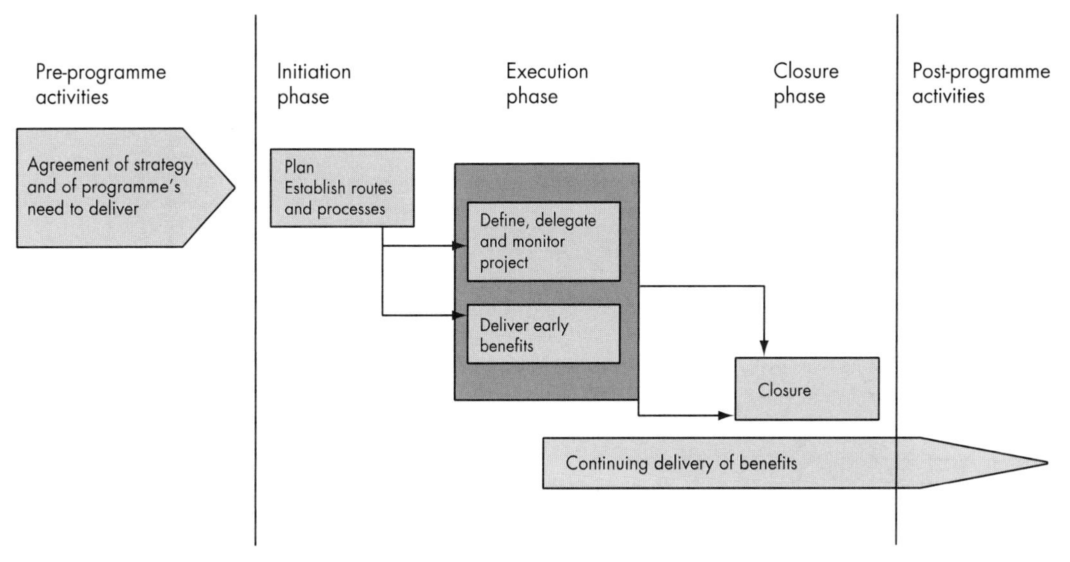

Figure 2.1 *Key stages in the programme life cycle*

delivered from early tranches. It is often impossible to define all projects in the initiation phase, so that phases often overlap.

It should be noted that not only are a wide range of terms used to describe these stages, but organisations in industry and the public sector may use additional stages to those proposed here. For example, 'initiation' is sometimes known as the 'definition' or 'establishing' phase. The fundamental principles remain the same.

Gateway reviews are recommended between these phases, and many organisations require them. These reviews ensure that positive business decisions are taken to continue with the programme in its current direction, change direction or abandon the programme completely. Also, gateway reviews provide ideal opportunities to ensure that governance directives are being followed.[7]

Prior to the start of any programme, there is usually a long, ill-defined and often uncertain and confused set of activities that eventually convince the organisation that it should invest funds in a programme. In theory, all organisations should have clearly defined and agreed strategies, the implementation of which requires the initiation of programmes. In practice, many organisations have unclear or poorly defined strategies. Even when the strategy is clear and agreed, there may be major differences of opinion about how it should be achieved, and therefore much debate and argument about which programmes to initiate and the resources to be committed. Frequently it will be necessary to conduct 'discovery' or 'pilot' projects, to undertake feasibility studies or to create proof-of-concept systems in order to help reach agreement.

However, at some point a decision is made to proceed and an instruction issued to start work. Such an instruction is sometimes referred to as a programme mandate. Once a decision has been made, the programme can be initiated. In this phase the governance arrangements will be established, key members of the programme management team appointed, and detailed plans and business cases defined. Approval to proceed to the execution phase will be requested through a submission to the programme board (see section 4). This submission may take the form of a programme initiation document or programme definition document.

[7] See also section 3.5 for further information on such reviews.

The execution phase follows the approval given by the programme board to proceed with the work, and is the longest and most costly phase of the programme. Component projects will be defined and initiated by the programme team and delegated to individual project teams. The projects will be monitored by the programme team so that out-of-tolerance variations can be recognised and managed. Where possible, early projects will be planned (in tranches) so as to create the potential for early benefits, and the realisation of benefits may start whilst other, longer-term projects are still being initiated. It should also be noted that often programmes start with business cases uncertain or incomplete, and that initial projects may be pilots or feasibility studies designed to provide the information and knowledge needed to produce a comprehensive and definitive business case.

The execution phase frequently includes projects that prepare the organisation to adapt to the changed or new capabilities that are being delivered. Aspects such as training, business process re-engineering and communications commonly involve business change management (see section 4) working in parallel with the programme management. A key role in the execution phase of a programme is to maximise the value gained from the time invested by the (normally) scarce resources. A programme team can take an active role in the allocation of resources across projects and minimise unnecessary duplication of work.

When all projects have completed and the new capabilities have been handed over to and accepted by the organisation, the closure phase is reached. These capabilities may now be used by the organisation to deliver improvements, and these are quantified through the measurement of benefits. The programme management team will encourage, support and monitor the delivery of benefits within the organisation's ongoing operations.

Frequently the delivery of improvements and the resultant realisation of benefits requires the use of new facilities as part of business-as-usual activities. As a result, benefit realisation activities may start soon after the completion of initial projects and continue long after the programme has been closed.

The three main stages of a programme are discussed in more detail in the following sections. Further information on the delivery of benefits can be found in section 3.2.

2.2 PROGRAMME INITIATION PHASE

2.2.1 Purpose and inputs

The objective of this phase is to properly establish a programme so that it will be possible to effectively co-ordinate its component projects and therefore deliver the programme's objectives.

Typically a mandate that initiates a programme would provide guidance on the:

- benefits expected;
- prioritisation of the programme;
- major risks and constraints associated with the programme;
- key milestones and any related programmes or projects that might be affected;
- results of any preceding 'discovery' or 'feasibility' project or similar preparatory work;
- budget available for the initiation phase, plus an outline estimate of time, cost and benefits of the overall programme;
- requirements for governance and reporting to the organisation's senior management;
- assumptions on which the programme is based;
- terms of reference for the programme manager.

2.2.2 Key activities

In the initiation phase the programme management team will be appointed, programme plans and management processes defined, and governance arrangements, such as a programme board, established. The end result will be the production of a formal justification for proceeding with the execution of the programme, typically in the form of a programme initiation document.

Typical key measures are listed below.

1 *Appoint the programme management team.* It will normally be necessary to appoint individuals to adopt key programme management roles,

such as the programme manager. He or she will then appoint key members of the programme team, such as the programme management office manager. These key team members should familiarise themselves with the programme, and its objectives and structure.

2 *Structure the programme and identify component projects.* Any early activity will be defining the overall structure of the programme and the identification of the component projects that will be required to deliver the programme's objectives and the interdependencies between these projects. The programme team should be in a position to give project managers clear project scopes and ensure that project managers understand how their projects contribute to the overall objectives of the programme and interrelate to other projects.

   Figure 2.2 shows an example of a programme structure diagram taken from a programme delivery plan. Here, a consortium of contractors (the Partners) is being led by a prime contractor (WizzCo) to undertake a programme of work for ABC Ltd. The diagram shows the main groups at each level with interfaces between them.

   It should be stressed that programme management structures vary significantly across programmes and organisations. Amongst the component projects may be one or more focused exclusively on the organisation's transition from the old ways of working to the new, as described in section 1.7. Alternatively, several component projects may include transition work packages and activities so that the deliverables they will create are separately put into use.

   Further information on typical programme roles and responsibilities can be found in section 4.

3 *Establish risk, issue and other programme-wide processes.* The programme team should ensure that necessary common processes, such as those needed to escalate risks and issues from the project level to the programme level, are in place. Where necessary, it should also ensure that other common processes, including quality, audit and testing requirements and the ways in which projects will be co-ordinated and will report to the programme, are in place to report on the progress of projects throughout their own life cycles.

   Where third parties, such as external contractors, are going to be responsible for managing some of the component projects, this activity

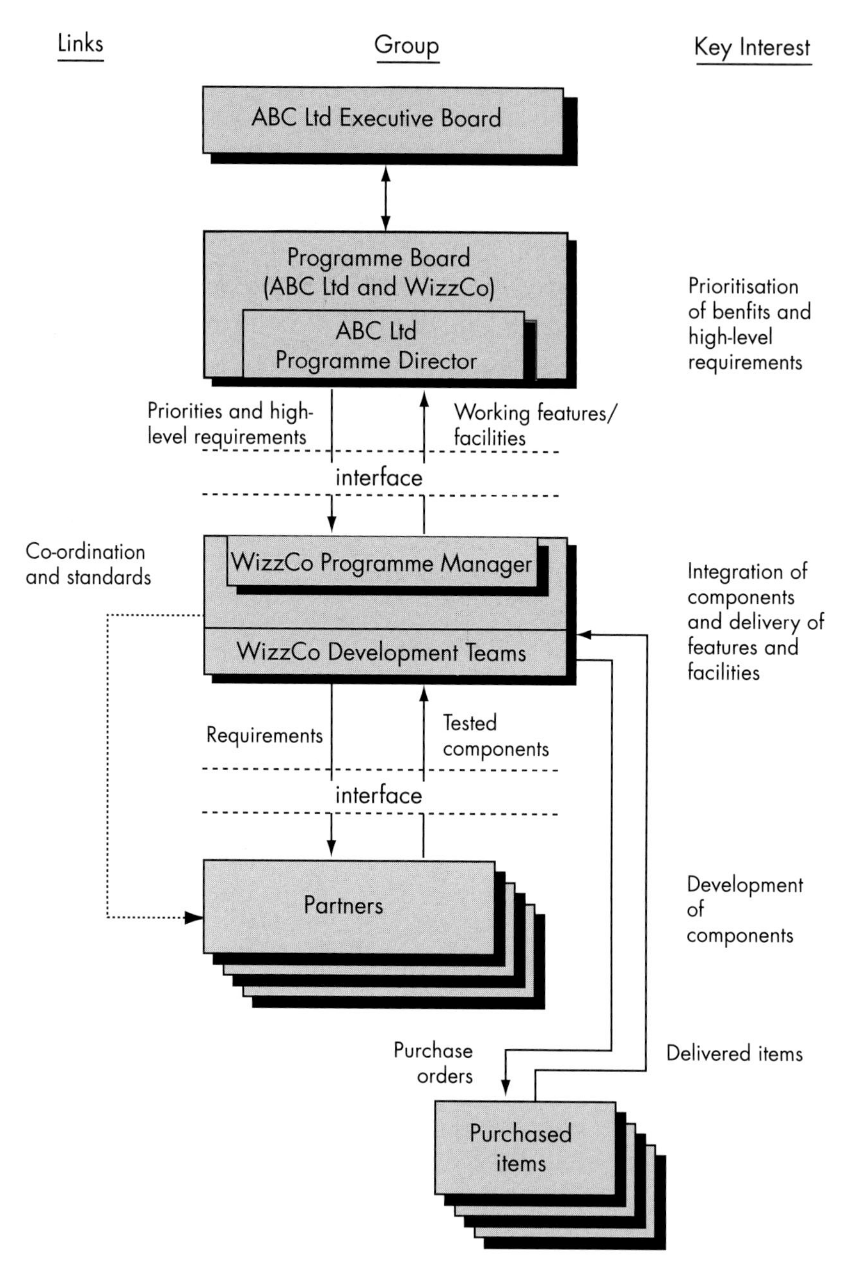

Figure 2.2 *Example of programme structure*

is particularly important. Only where there are standard and agreed processes, supported by appropriate mechanisms for auditing and enforcing conformance to these processes, can the programme management team be confident that they will be able to co-ordinate the different projects and interface their various deliverables.

4 *Establish programme governance.* It will usually be necessary to establish and define specific arrangements for programme governance. Typically these will involve appointing a programme board and defining its role, responsibilities and processes. These processes should be defined in writing so they can be followed efficiently.
5 *Agree programme assurance and audit arrangements.* Normally, the programme board will require a separate assurance and audit function within the programme, which reports directly to them rather than through the programme team. This is to ensure that the programme assurance function can provide an independent source of information to the board that is not filtered by the programme manager. Further guidance on the programme assurance function can be found in section 3.5.
6 *Identify training needs.* It is often the case that staff members are unfamiliar with the demands of programme management roles, including those involved with programme governance. Initial plans and budgets may therefore need to allow for appropriate, necessary training.
7 *Establish a programme management office (PMO).* A large programme can encompass massive amounts of administration involving planning, operational, financial, communication and HR issues. A PMO is normally essential to handle this administration and to provide the programme manager with the information he or she needs at all times to maintain control. Furthermore, the PMO can be a great help during the initiation phase. Further guidance on the role of the PMO can be found in section 3.6.
8 *Establish financial control and reporting arrangements.* It is important to set in place arrangements for financial control and reporting. Normally, each project will report at regular intervals to the programme team (typically through the PMO), which will accumulate the results and report on progress at a programme level.

Programmes are frequently major undertakings involving the expenditure of large sums of money. The financial arrangements will need to interface appropriately with those of the organisation.

9 *Identify any special management techniques.* Because of their size and complexity, programmes frequently need especial attention to key areas of management, including:
   - planning and control
   - benefit management
   - risk management
   - change management – i.e. the management of business or organisational change
   - stakeholder and communication management.

   Further guidance on each of these and other relevant techniques can be found in section 3.

10 *Prepare programme plans.* The programme plan and other programme-level documents, including the financial justifications and benefit delivery plans, should indicate how the programme will be organised, how projects will interface with the programme, and how benefits will be realised. These plans should also identify any special techniques or approaches that will be required within the programme, as described in section 2.2.3, and in section 3.

2.2.3 Phase outputs

The key output from the initiation stage should be a detailed business case justifying proceeding to the next phase. This document, often called a programme initiation document, should provide sufficient information to allow the programme board to make a rational decision that the likely investment will be justified by the benefits that will ultimately be achieved.

This document will often be accompanied by other key outputs, including:

- the programme plan, together with any necessary supporting plans or documents – including processes to be used throughout the programme;

- the project list – identifying the projects required to deliver the programme;
- budgets and financial plans;
- the initial programme risk register;
- the response plan;
- stakeholder analysis and the communication plan.

2.3 PROGRAMME EXECUTION PHASE

2.3.1 Purpose and inputs

The purpose of this phase is to effectively manage the programme so as to deliver what has been planned (including any organisational change and business benefits) in accordance with agreed timetables.

Because of the typical size and complexity of programmes, all component projects and deliverables may not be fully defined at the start: part of the execution phase involves identifying more precisely what is to be delivered and gaining agreement throughout the programme's life. Thus, maintaining an understanding of stakeholder expectations, and managing those expectations and ensuring that they are met, is usually vital to success.

Inputs to the programme execution phase will include the business case, identifying the benefits that are required, together with the other plans and processes prepared during the previous phase and recorded in the programme initiation document.

2.3.2 Key activities

This phase involves the successful management and execution of the programme. Once the programme is fully established, the component projects will be created to produce the required deliverables (including new services). These deliverables will be combined under the control of the programme to produce the new facilities or capabilities that can be exploited to realise benefits. In parallel with the development of capabilities is the preparation for their implementation. During this phase these projects will be co-ordinated and managed efficiently so as to successfully deliver all the desired outputs.

Changes to the organisation's strategy or business environment may result in significant alterations to the expected benefits, and thus require adjustment of the business case upon which the decision to proceed was made. Thus it is good practice to have major reviews of the programme's viability at key stages throughout the execution phase.

The principal activities involved in directing a programme are described below. These should all have been identified during programme initiation and be described within the programme plan or supporting documentation.

1 *Establish component projects.* The component projects themselves must be established and planned in detail, cross-checked and optimised for technical, managerial and commercial consistency. The individual project plans, time schedules and other supporting documentation should be reviewed (by the programme manager or by the PMO on his or her behalf) to ensure consistency and to identify dependencies and potential conflicts.

   For each project, a separate project manager should be appointed. Managing a project that is part of a larger programme is slightly different from managing a stand-alone project, and appropriate terms of reference, identifying different reporting structures and escalation procedures, should be provided, agreed and signed off.

2 *Co-ordinate project plans and schedules.* Although each component project will develop its own project plan and attendant time schedule, successful programme management requires these to be co-ordinated. This requires all interdependencies to be identified, and then the individual plans and schedules adjusted to achieve the best possible compromise.

   Once a consolidated schedule has been agreed, it is likely to need constant adjustment as progress and changes are notified to ensure that any delay in one component project is accommodated, thus avoiding 'knock on' delays to other projects and thus to the delivery of the desired programme outputs.

Through the co-ordinated management of the component projects, the programme will deliver the required outputs and outcomes in the most cost effective manner.

3 *Manage risks and issues.* As with individual projects, rigorous risk- and issue management is essential. Typically, each project will maintain its own risk register and manage its own risks, but will escalate to a programme-level risk register those risks that are beyond its control.
4 *Manage stakeholders and communications.* Whilst co-ordinating and managing the component projects within the programme, the programme team must also develop relationships with the customer and with other stakeholders.

In complex environments, such as those relating to programmes, stakeholder management cannot be undertaken in isolation from other management activities. Instead, it must be part of the overall approach to gaining and then maintaining the support and co-operation of all stakeholders, and must be co-ordinated with related activities such as governing the programme, communicating issues and progress, and mutual problem solving.

5 *Report progress.* At agreed intervals (typically monthly), consolidated programme status reports should be produced. The scope and coverage of these will have been defined during the programme initiation phase according to the needs of the programme board.

Typically such reporting will include programme and financial status reports – based upon information provided by the component projects. These should give a concise 'snapshot' of the status of the programme, identifying progress against milestones and any major new risks, and concentrating on exceptions and departures from agreed plans. A key part of such a report will be the progress towards the delivery of benefits. The financial status reports should give a summary of the consolidated costs, revenues and working capital of the programme.

Producing these reports will require each project to prepare its own individual reports and then forward these to the programme for consolidation, normally by the PMO. To ensure that meaningful results can be prepared by the agreed dates, it will normally be necessary for the

PMO to define and enforce a standard reporting timetable and to provide templates in which the managers of component projects can record their information.

6 *Review and update programme plans.* The programme plan and project plans and schedules are likely to need regular updates as a result of:
   - activities that were known to be required, such as the initiation of a new project, but which could not be planned in detail at the start of the programme;
   - changes to the plans and schedules of existing projects as a result of delay or unexpected problems or difficulties;
   - alterations to the scope, content or composition of the programme as a result of change requests.

   Major updates should be discussed with and agreed by the programme board.

7 *Hold progress reviews.* It is good practice for the programme board to conduct progress reviews at critical points in the life of the programme, such as the end of a design or development stage or immediately before commencing implementation. Such review points are usually identified at programme start-up and should be identified in the programme plan, a part of the programme initiation document.

Of course, such reviews are time-consuming, and thus adequate budgets and resources should be provided for the programme management team to prepare for them. These reviews may take the form of 'gateways' where a formal go/no-go decision is taken about the subsequent stage.

2.3.3 Phase outputs

The principal outputs from the execution phase will be the new facilities and capabilities that the programme is intended to create, plus any benefits that were delivered during the programme life cycle.

The programme will create a range of detailed outputs, such as revised plans, programme and project progress reports, risk and issue logs, internal invoices, test and training records, and so on. It will usually be the responsibility of the PMO to ensure that these are adequately filed.

2.4 PROGRAMME CLOSURE PHASE

Upon completion of the programme, there should be a closure review, based upon that undertaken at the completion of a project through the process. The purpose of this review will be to verify, amongst other things, that:

- all deliverables and capabilities have been delivered and handed over successfully;
- all projects have completed their own individual project closure;
- all necessary records are in place;
- all customer and supplier invoices have been processed;
- lessons have been learned and other valuable knowledge, including an up-to-date programme summary, has been captured.

The point at which such a review takes place will depend upon the nature of the programme. For example, if the output of the programme is a new facility, closure is likely to occur as soon as the last project has completed and the facility has been handed over to the line management. Alternatively, where the programme is required to deliver a complex range of business benefits, there may need to be a period of use by the customer of the new facilities or capabilities before the benefits can be realised. In these circumstances, there may need to be a delay between handover and the final closure review.

With a large and complex programme, the closure review may be staged and may involve considerable effort, so time and cost should be allowed within budgets and schedules.

3
Supporting programme management techniques

3.1 PROGRAMME PLANNING AND CONTROL

The main objective of programme planning and control is to maintain high-level plans of the component projects so as to identify interdependencies between those projects. These dependencies may take a number of forms, as summarised below.

- Logical interdependence occurs where a task in one project plan cannot proceed without a deliverable from a task within another project plan. Whilst each project manager should maintain a plan for his or her own project, the programme plan must acknowledge and monitor these logical interdependencies.
- Benefit interdependence occurs where a benefit can be achieved only as a result of the outputs of a number of projects. Therefore the effect of one of the projects leading to that benefit being delayed, changed in terms of direction or scope, or cancelled must be tracked and monitored so that the impact on benefit delivery can be fully understood.
- Resource interdependence is a feature of nearly all programmes, as key resources are shared across the component projects. The requirement to model, predict and monitor the demands and availability of resources across the programme falls within the remit of the programme planner.

Programme planners also assist and support project planners with the objectives of ensuring that projects are planned in accordance with the programme's standards and with sufficient consistency to allow for the regular integration of plans for programme reporting.

3.2 BENEFIT MANAGEMENT

The business discipline of defining, agreeing, planning, measuring and reporting on benefits is referred to as benefit management. The relationship between projects, programmes and benefit management is frequently quite complex.

Rather than delivering benefits, projects create deliverables such as a working IT network, a tested software application, a pool of trained staff, and so on. These projects must be managed by project managers who are familiar with the methodologies and techniques needed to produce those deliverables. The effectiveness of the project manager is determined by the extent to which he or she provides the required deliverables on time, on budget and to specification.

Programmes themselves rarely deliver benefits directly, but by combining projects and their deliverables, they create the capabilities and facilities that will enable the desired benefits to be achieved. It is then up to the line management to ensure that they are used properly so as to deliver the desired benefits. Programme management teams can support, measure and monitor benefit delivery but cannot, without line management's input, deliver benefits.

The benefit management process tries to ensure that the new facilities and capabilities created are utilised fully by monitoring, measuring and facilitating the delivery of the benefits.

Benefits can take many forms. Whilst financial benefits are clear and easy to understand, non-financial and intangible benefits are often just as valuable, though often harder to measure. Table 3.1 is an extract from a benefit plan showing three different benefits that are expected from a programme to create and implement a new web-enabled procurement system. In this extract:

- benefit 1, improved management control, is an intangible benefit – i.e. it cannot sensibly be measured in any meaningful way;
- benefit 2, reduced handling costs, is a tangible, financial benefit which can be measured and expressed in monetary terms – sometimes such benefits are also referred to as 'bankable';

Table 3.1 *Extract from a benefit plan listing benefits to be expected from a programme*

Benefit	Component project	Description of benefit	How delivered	When realised	Category of benefit	Benefiting stakeholder
1. Improved management control	Implementation of an e-procurement system, including staff changes and retraining	Identify amount of money spent with each supplier – better management control	All purchases made through raising formal purchase orders, which are all summarised in the same database	Nine months from start	Internal management	Purchasing department
2. Reduced handling costs	Implementation of an e-procurement system, including staff changes and retraining	Reduced order-handling costs – 3% financial saving expected	All purchases made through Internet by requisitioner; no additional manual processing by purchasing department required	Nine months from start	Financial	Purchasing department
3. Reduced complaints	New warehouse, including construction, IT support, staff changes and training	Better customer service – 10% reduction in number of complaints	New warehouse will allow products to be stored separately in unique bins or pallets, preventing loss or mix up through the use of shared bins	Twelve months from start	Customer service	Sales department

- benefit 3, reduced complaints, is also tangible, but non-financial – i.e. it can meaningfully be measured in non-financial terms.

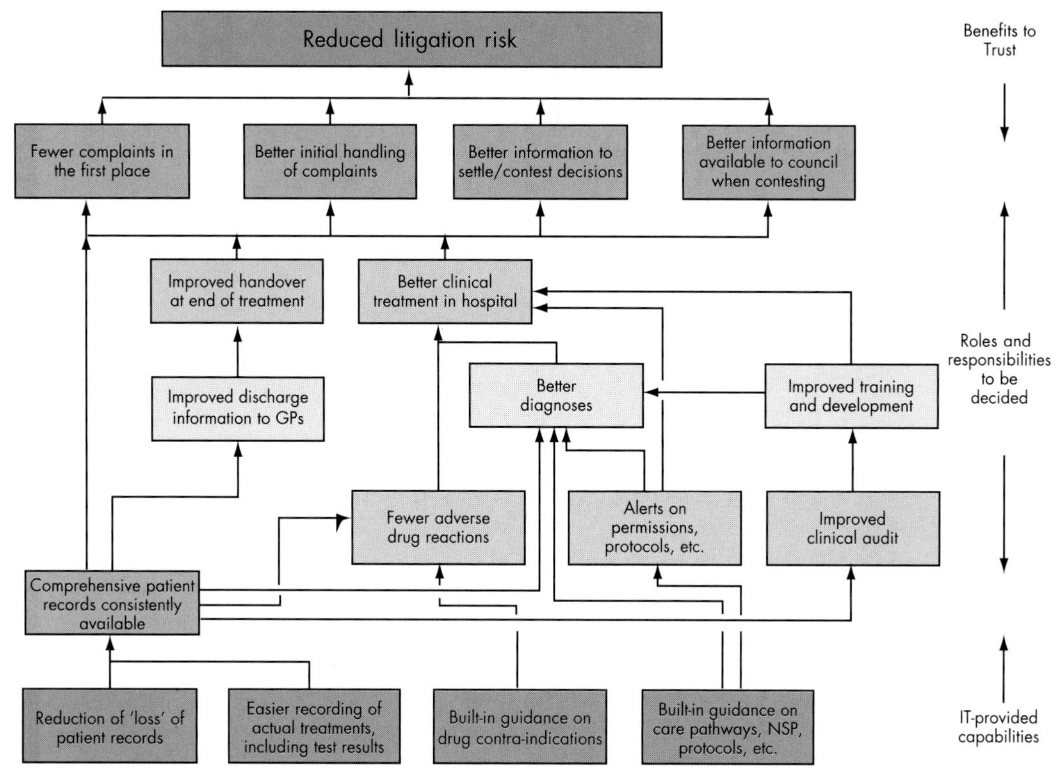

Figure 3.1 *Example of a benefit map*

Many programme management experts argue that all benefits must be measurable and that intangible benefits should be ignored.

The relationships between deliverables, facilities/capabilities and benefits are usually complex, and part of the benefit management process involves gaining an understanding of these relationships. One approach that is frequently used is to display them in a graphical form as a benefits map, as shown in Figure 3.1. This map shows how providing new IT facilities within a hospital might be combined and used so as to reduce the litigation risk (i.e. the risk of being sued for negligence or medical malpractice).

3.3 PROGRAMME RISK MANAGEMENT

Managing risk at a programme level requires that risks are also managed effectively at the project level. Every project should accept the responsibility for managing risks that impact the delivery of the project's goals. In addition, there are risks within a project that may impact other projects or the programme as a whole, and these should be escalated to the programme level. In addition, there may be risks that affect no single project but which could have a major impact on the programme and its delivery of benefits. Both escalated and programme-level risks should be managed within the programme. A role should exist within the programme team to consider, evaluate and monitor risks that may have an effect at the programme level, and whoever has this responsibility should ensure that:

- risks at the programme level are identified, evaluated and managed – this frequently involves benefit risks, i.e. risks likely to impact the delivery of benefits;
- each project has a defined risk management process and that risks are appropriately managed;
- there is a process for identifying and escalating appropriate risks from the project level to the programme level;
- there is a process for seeking risks that relate to or are shared by multiple projects so that a programme-level risk management may be considered.

Without effective programme-level risk management, the benefit risks may be poorly understood or ignored. For example, it is not unusual to de-scope an IT project in an attempt to deliver on time. The impact of this reduced functionality may well be that the project is perceived to end on time when the new function is handed over to the user after acceptance testing in accordance with the schedule. However, if the reduction in functionality no longer allows the expected benefits to be delivered, the organisation's investment will have been wasted.

Programme-level risk management should also be integrated with the organisation's strategic risk management. For example, where multiple change programmes are being run, an alteration to the priorities of one programme may impose a benefit risk to another. So, just as conflicts between the objectives of individual projects should be escalated to the programme management team for management and resolution, conflicts between programmes should be escalated to the organisation's main board or risk committee for management and resolution.

3.4 STAKEHOLDER MANAGEMENT AND COMMUNICATIONS PLANNING

Stakeholder management (sometimes also referred to as stakeholder engagement) is the means by which the programme team manage the relationships with the wide range of people and organisations involved in and impacted by a programme.

In the broadest sense, anyone who is involved in, impacts, is impacted by, or influences the programme or its environment can be regarded as a stakeholder. Part of stakeholder analysis involves gaining an insight into the stakeholder's views of the programme and its benefits, and how these views might be made more positive. Table 3.2 shows an example of stakeholder analysis. A particularly important group for many programmes will be the line management who, by using the facilities and processes created by the programme, will actually realise the expected benefits. At this stage some programme management teams will include additional columns covering actions required and expected outcomes to

Table 3.2 *Extract from part of a stakeholder analysis*

Stakeholder	Role	Influence	Analysis
Reginald Fryant	Chief executive (CEO)	Strong	Neutral
Jon Siligo	Chief information officer (CIO)	Very strong	Ally
Wendy Smith	Finance director	Medium	Ally
Frank Roosevelt	Non-executive director	Weak	Opposer
Elizabeth Woods	Director, customer operations	Medium	Opposer
Patrick Bragg	Staff association representative	Strong	Opposer
Ron Jones	Financial journalist	Medium/strong	Opposer
Kannan Sami	Investment analyst for major stockholder	Weak	Neutral

Table 3.3 *Extract from a communications plan*

Objectives	Audience	Key messages	Timing	Media
Win support for business case	All stakeholders	Clarify benefits to company Risks of not off-shoring	Now and week before board meeting	Briefing paper Direct briefing by finance and programme director
	CEO CIO Finance director Non-executive director Investment analyst for major stockholder	Impact on margins Potential positive impact on share price		Briefing paper Market analysis
Neutralise opposition	Financial journalist Director, customer operations	Programme is enabler for more investment in customer services	Now and week before board meeting	Press release and interview
			Now and weekly	Proposition paper and meeting(s) to ask for suggestions
	Director, customer operations	New international role	Now	Proposition paper and meeting(s) to ask for suggestions
	Staff association representative	Redeployment opportunities, no enforced redundancy and generous voluntary redundancy	Now and monthly to board meeting	HR briefings

summarise the management of stakeholders and especially 'Opposers' or 'Neutrals'.

A large part of managing stakeholders is achieved through communications planning, which determines the amount, manner and purpose of the communication to be sent to and received from each stakeholder. The resultant communications plan should identify the stakeholders and the messages the programme team wish to send to those stakeholders, as well as the media and timing of those messages. Table 3.3 shows an extract from a communications plan.

3.5 PROGRAMME ASSURANCE

Establishing an effective programme assurance regime to determine quality requirements and ensure that they are met is vital to the success of every programme. Without such a regime, it is unlikely that the long-term objectives of the programme will be achieved. Moreover, this regime must be much more comprehensive than that which applies within individual projects. It must ensure not only that the deliverables of all projects are of adequate quality, but that they can be combined with others to create the necessary capabilities which will deliver the long-term objectives of the programme.

The objective of a programme assurance function is to implement a programme quality environment that ensures that quality requirements are identified, understood and fulfilled through all parts of the programme and through all programme activities. These activities include those of the programme management function, as well as all relevant activities within the component projects.

Managing quality at a programme level implies ensuring that quality is managed effectively at the project level. This is particularly the case where different organisations are undertaking different projects, as applies within a consortium, for example. In these circumstances it is unusual for a single project management methodology to be mandated for every project, since many organisations have their own distinct approach to quality, and requiring changes to this could create more quality problems than would be resolved through consistency. However, what is important in these circumstances is that:

- a common interface is established between the component projects and the programme;
- the systems operating within the component projects meet the requirements of the programme, as determined by the programme assurance function;
- the programme assurance function has adequate visibility of what goes on within the component projects – this may require regular quality reporting plus audit arrangements so that the programme assurance

function can confirm that what is reported is correct and truly representative of project reality.

It is important when using external contractors to manage component projects to ensure that quality requirements, the right to audit and an obligation to respond with corrective action plans to any audit observation is included within the formal contracts.

As mentioned in section 2.2.2, the programme assurance function should have an independent line of communication with the programme board.

Gateway reviews are sometimes used to support programme assurance. A gateway review can be carried out at both the project and programme level at key points in the programme life cycle. It is a formal review to confirm both the validity of the programme and the quality of its management and control. Gateway reviews may be carried out by teams external to the programme and external to the organisation: for example, the Office of Government Commerce (OGC) offers a team of expert gateway reviewers able to report on programmes at key points in their life cycles.

3.6 PROGRAMME MANAGEMENT OFFICE

The programme management office (PMO) is the hub of a programme and performs a collection of functions that provide valuable services to programmes. Most programmes would normally benefit from the services of a PMO, and the programme office manager frequently deputises for the programme manager. Typical roles and responsibilities are summarised below.

1 *Strategic support.* Many programmes exist to implement part of the organisation's strategy. A PMO can support this obligation by ensuring that programme teams have access to the correct versions of strategic information, as relevant to their scope, and informing programmes when strategy changes (or when significant change is pending). The

PMO can direct lessons learned and other information to the senior management, so they can understand whether strategic initiatives are, or have been, successful.

The PMO can help to locate programmes that may be in danger of implementing change not required by the strategy, and can make sure programmes are aware of each other's strategic contribution, to avoid conflict, make teams aware of cross-programme dependencies and promote synergistic approaches through cross-programme collaboration.

2 *Programme support.* The PMO provides management support to the programme manager, including:
   - expertise on programme management methods, processes and standards which are used within the organisation;
   - support to the programme assurance function audit/assurance – for example, to monitor corrective actions following audit observations;
   - custodianship of programme information, to include master/baseline copies of documentation, configuration management, and version and change control;
   - provision and analysis of programme information, via reporting or other methods;
   - publication of updated progress information, acting as an information centre on all projects and programmes and producing regular update reports;
   - advice to the programme manager to support his or her decision making;
   - assistance with related techniques, investment appraisal, prioritisation, non-financial benefit measurement methods and others – this may sometimes be provided via an alliance between the PMO and other organisational functions, such as finance;
   - maintenance of baselines including budgets, schedules and benefit plans, including the updating of these baselines when change control procedures make this necessary.
3 *Resourcing.* Programmes and their projects require resources on both a full- or part-time basis. These resources can be people, where the

requirement will be for specific skills, or physical resources where the requirement will be for specific functions (e.g. testing facilities).

A PMO can assist with resource management by:

- maintaining a database of human resources – their skills/attributes, location, availability, contact details and managerial responsibility for the resource;
- monitoring other programmes, projects and business-as-usual activities which will impact the ability of a programme to deliver;
- maintaining relationships with external organisations that can supply resources, such as contract agencies for staff, organisations that rent plant and equipment, and agencies that let building space;
- supporting the management of the portfolio of programmes and projects by assisting with allocating resources to programmes and projects – where resources are employed on more than one programme or project and business-as-usual activities at the same time, some form of matrix management might be employed.

PMOs do not need to exist as one single functional unit. They can be:

- dedicated PMOs, each serving a particular programme – this is the clearest pattern and ensures that the programme manager receives full support;
- shared units, serving all programmes and projects within the organisation;
- shared units, but serving a particular region, division or geography – this is often the pattern where the office needs to fully understand the local culture, legislation and business practice.

PMOs should not be seen in a simply administrative support role. Some PMOs take a proactive role in the management of programmes, including:

- project initiation
- project manager selection and appointment
- integration across projects

- project closure
- internal and external charging
- change control.

3.7 OTHER PROGRAMME PROCESSES AND TECHNIQUES

Many of the processes and techniques used within programmes are similar to those used within projects, such as leadership, team building and staff motivation. However, there may be a difference of emphasis when operating these processes or techniques at a programme level, as compared to at a project level. For example, change control within a project generally focuses on changes to required deliverables and the resultant impact on project cost, delivery schedule and quality. By contrast, change control at the programme level focuses on changes to the organisation's business priorities, and the resultant impact on expected benefits and the overall business case of the programme.

Thus, whilst the concepts, processes, forms and tools may be similar, the outputs may vary greatly at the two levels.

4
Roles and responsibilities in programme management

Within a complex organisation such as a programme, it is vital that all concerned understand their roles and responsibilities and those of the people that they must interface with. Figure 4.1 shows a typical programme organisation chart, identifying the key positions.

For each of the posts identified in the chart, there should be clear terms of reference. A summary of those posts shown on the chart that are most relevant to programme management is given below.

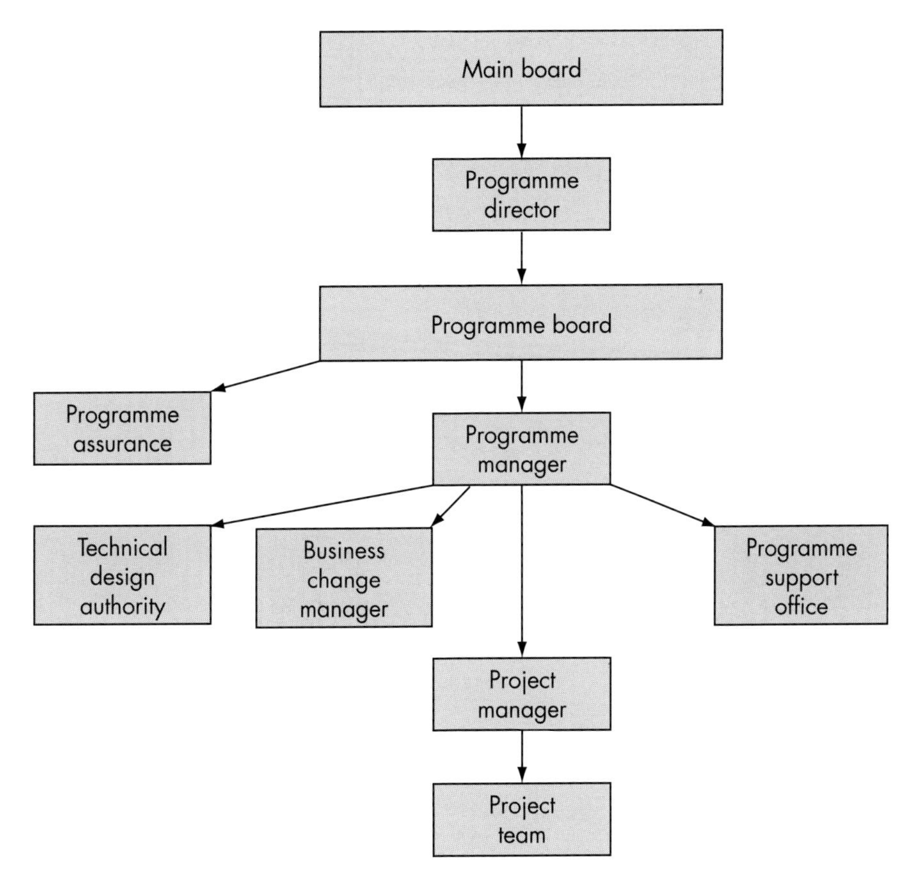

Figure 4.1 *Typical programme organisation chart*

1 *Programme director, programme sponsor or SRO*. For each programme there should be a senior executive allocated to direct the initiative in line with corporate strategy and to provide high-level sponsorship. Typically the programme director would be a member of the main board, and thus familiar with the organisation's strategy and high-level priorities. The responsibility for making the investment decision and providing senior endorsement for the programme rests with this individual, but this role is a delivery role. In some organisations this role is referred to as the senior responsible owner (SRO) and in others the programme sponsor.
2 *Programme board*. This is the group or committee responsible for the direction-setting and leadership of a programme. It usually includes representatives of key stakeholder groups, partners or investors. Its role is to verify that the programme remains on course to deliver the expected benefits and to advise on any changes to plans, scope or programme priorities. The programme director would typically chair such a board.
3 *Programme manager*. This person carries out day-to-day management of the programme and supervises the component projects. He or she is responsible for ensuring that the programme progresses in accordance with the policies and priorities identified by the programme board.
4 *Programme assurance*. The programme director and programme board members often have business-as-usual responsibilities and can give only a limited amount of time and attention to the programme. As a result, they rely heavily on information provided by the other members of the programme team. Whilst they may get comfort from the receipt of regular reports and communications, they need to have independent assurance that everything is going as the programme manager describes. This is a key role of the programme assurance function.
5 *Technical design authority*. Where the component projects involve an element of technical design, it is essential to have a common approach, to ensure not only that the various technical products will interface correctly with each other, but also that they do not compromise the existing technical infrastructure. The technical design authority is responsible for devising this common approach and ensuring that the